MOLECULES AT

John Emsley lectured in chemistry for twenty-five years in the University of London, and he is the author of over a hundred research papers. He is now Science Writer in Residence in the Chemistry Department at the University of Cambridge. Dr Emsley's 'Molecule of the Month' column for *The Independent*, which ran from 1990 to 1996, brought home to a wide readership how chemistry impinges on every aspect of our daily lives. In 1993 he received a Glaxo Award for science writing, and in 1994 won the Chemical Industries Association's President's Award for science communication. John Emsley's much praised book *The Consumer's Good Chemical Guide* won the Rhône-Poulenc Science Book Price in 1995.

MOLECULES AT AN EXHIBITION

Portraits of intriguing materials in everyday life

JOHN EMSLEY

OXFORD
UNIVERSITY PRESS

OXFORD
UNIVERSITY PRESS

Great Clarendon Street, Oxford OX2 6DP

Oxford University Press is a department of the University of Oxford
and furthers the University's aim of excellence in research, scholarship,
and education by publishing worldwide in

Oxford New York
Athens Auckland Bangkok Bogotá Buenos Aires Calcutta
Cape Town Chennai Dar es Salaam Delhi Florence Hong Kong Istanbul
Karachi Kuala Lumpur Madrid Melbourne Mexico City Mumbai
Nairobi Paris São Paulo Singapore Taipei Tokyo Toronto Warsaw

and associated companies in Berlin Ibadan

Oxford is a registered trade mark of Oxford University Press
in the UK and in certain other countries

Published in the United States
by Oxford University Press Inc., New York

First published 1998

First issued as an Oxford University Press paperback 1999

British Library Cataloguing in Publication Data
(Data available)

Library of Congress Cataloging in Publication Data
(Data available)

ISBN 0-19-286206-5

Typeset by Footnote Graphics, Warminster, Wilts

5

Printed in Great Britain by
Clays Ltd, St Ives plc

CONTENTS

GALLERY 3

Starting lives, saving lives, screwing up lives

An exhibition of molecules that can help and harm the young

GALLERY 4

Home, sweet home

An exhibition of detergents, dangers, delights and delusions

GALLERY 5

Material progress and immaterial observations

An exhibition of molecules that make life a little easier

GALLERY 6

Landscape room: environmental cons, concerns and comments

An exhibition of molecules that stalk the world

GALLERY 7

We're on the road to nowhere

An exhibition of molecules to transport us

ACKNOWLEDGEMENTS

IN WRITING this book I have drawn upon many sources and talked with many people. To track many of them down I have often relied on the Media Resource Service which is part of the Novartis Foundation, London, and is run by Chris Langley, Jan Pieter Emans and Janice Leeming.

Much of the work I do is commissioned by the editors and staff of newspapers and magazines, and some have indeed become friends. They are Tim Clark, Dick Fifield, Victoria Hook, Ron Kirby, Chris King, David Pendlebury, Cath O'Driscoll, Tim Radford, Karen Richardson, Gill Rosson, Rick Stevenson and Tom Wilkie.

For moral support, and useful discussions about the meaning of life and the role of chemistry in it, I am indebted to present and past colleagues in the Chemistry Department at Imperial College, especially to Tony Barrett, Jack Barrett, Don Craig, Sue Gibson, Bill Griffith, Tim Jones, Steve Ley, David Phillips, Garry Rumbles and the late Geoffrey Wilkinson.

The idea of an exhibition of molecules can be traced to my friendship with Alfred Bader, who has been successful both as a chemist and as an art collector.

Molecules at an Exhibition is also a tribute to the continued support of my publisher, Michael Rodgers, the copy-editing skills of Jane Gregory, and of friends and family who kindly read first drafts, telling me when they did not understand what I had written. Norman Greenwood provided very welcome help at the manuscript stage by pointing out errors and suggesting useful additions. For any lapses in content and style thereafter, and for any scientific errors which have still escaped detection, I take full responsibility.

INTRODUCTION

THIS BOOK is a development of articles which I wrote originally for newspapers, magazines and various company publications. Some of the topics covered in *Molecules at an Exhibition* are based on subjects that appeared in a regular column I wrote for *The Independent* newspaper, called 'Molecule of the Month' which ran for six years from 1990 to 1996. The more light-hearted ones are developed from pieces in the 'Radicals' column of the magazine *Chemistry in Britain*, which is the monthly journal sent to members of the Royal Society of Chemistry. Several portraits in *Molecules at an Exhibition* are from my private collection. They are subjects that have simply taken my fancy and I am writing about them for the first time.

Writing for newspapers and magazines means working within the confines of deadlines and fixed numbers of words. These constraints may be good at focusing the mind, but they have their limitations. They mean that a lot of background material, interesting side issues, historical perspectives, and my own personal views have to be omitted. Writing a book enables me to put all these in, and it also enables me to provide a broader perspective on molecules that feature in the news for a few days but then are rarely heard of again.

The title *Molecules at an Exhibition* is easily explained. It emphasizes that this is a personal collection of chemicals that I find particularly interesting. Each portrait is complete in itself and I have tried to make the collection as varied as possible. I have grouped them into eight galleries, each with a common theme. You may think I have placed some of the portraits in the wrong collection, and had I put this exhibition together a few years ago some of the portraits might well have been hung differently. Molecules we once regarded as obnoxious and dangerous turned out to be essential to the workings of the human body. Selenium and nitric oxide come into this category, so instead of finding themselves in the rogues' gallery, one is in Gallery 1, which is devoted to unusual chemicals in the food we eat, and the other in Gallery 3, which is devoted to sexually important molecules.

Millions of different molecules have been made since chemistry began in the eighteenth century. A select few, perhaps one in a thousand, turned out to be important, and it is these which play a part in our everyday lives. The majority of new molecules have had but a brief existence. They were made or discovered; they were examined and their properties recorded; they were reported in scientific journals or mentioned in patents. And that was the end of the matter. They might still be stored somewhere, but most have now been lost. (Dr Alfred Bader, the founder of the Sigma Aldrich Chemical Company, made a point of buying many of them from their original discoverers and storing them for future use by chemists.)

The same could be said of paintings, whose number no doubt has exceeded many times the number of known molecules. Most of them have suffered the same fate of neglect and been lost, and yet the important ones are still around and these are the ones you expect to see when you visit an art gallery or exhibition. And yet when you visit a gallery or exhibition you might discover some by little known artists that are just as fascinating.

It is in this frame of mind that I would like you to enter this Exhibition and tour its portrait galleries, rogues' gallery and landscape room. We have portraits devoted to culinary delights, health, plastics, domestic skills and transport. You will already have heard about many of the molecules on display, but I hope that you will learn something more about them now you have a chance to inspect them more closely. A few may be unfamiliar, but even these will have some role to play in your everyday life, and I hope that you will find examining their portraits too an enriching and entertaining experience.

You should not need a degree in fine arts to take pleasure when viewing great paintings, one in music to appreciate a symphony concert, one in media studies to enjoy a movie, nor one in literature to be enthralled by a good book. Nor do you need a degree in chemistry to read and understand *Molecules at an Exhibition*. Language is still the primary medium of communication but it can also be a barrier, and scientific language can be one of the most effective barriers to understanding. I hope this is not the case in this book, and for this reason I have not included any chemical formulas, equations or molecular diagrams with the portraits. If you want to learn more about a particular molecule, there is a list of other books you can consult at the end.

Come on in! There are eight galleries to explore, so wander at will. Each gallery contains about a dozen portraits—and lots of surprises.

A QUICK GUIDE TO THINGS LARGE AND SMALL

I WEIGH 13 stone in the UK, 182 pounds in the USA, and 83 kilograms in continental Europe. I am six feet tall in the UK and the USA, but 1.83 metres in continental Europe. If I buy a ton of sand in the UK I would get more than in the USA (2240 pounds as against 2000 pounds), but about the same as a tonne in continental Europe, which is 1000 kilograms (2205 pounds). There are many other examples I could quote where the same quantity can be expressed in Imperial, US and metric units which give very different numbers.

Science has its own scale of measuring things, called SI units (short for Système International), which is derived from the metric system. This enables us to talk about very small quantities, and is essential for chemistry because this science is about a world we cannot see, a world of atoms, molecules and minute traces. These have also to be measured, and in quantities that are based on the larger weights and measures. If you are unfamiliar with these, or with the SI system, then here are some examples that may help you.

◆

The average adult human weighs 11 stone, which is 154 pounds or 70 kilograms (kg). From now on it is best to forget the first two and concentrate on the third of these:

70 kg is the same as 70 thousand **grams** (g)
70 million **milligrams** (mg)
70 billion **micrograms** (mcg)

These smaller measures of weight can be visualized separately:

A gram is about the weight of a peanut.
A milligram is about the weight of a grain of sand.
A microgram is about the weight of a fleck of dust.

Many molecules are found in only tiny amounts in soils, in water, in the air, or in the human body. When we talk about these molecules we need to express the amount as a fraction of the whole.

Small quantities are often described as fractions of a percent. For example, 0.1% is one gram in a kilogram.

Tiny quantities are often described in parts per million (ppm), which is equivalent to one gram in a million, in other words one gram in a tonne (a tonne is a million grams).

Incredibly small quantities are described in parts per billion (ppb), which is equivalent to one milligram in a tonne.

Unbelievably tiny quantities are described in parts per trillion (ppt), which is equivalent to one microgram in a tonne*.

Some chemicals are produced on a large scale, and there we need the more familiar units of weight and volumes, which are kilograms and litres. In the metric system we can link these two units very easily when we are talking about water because one litre of water weighs one kilogram.

Very large quantities are measured in tonnes. A tank which is one metre (or about a yard) square and one metre high can contain one cubic metre of water and it weighs exactly a tonne. A cubic metre is equivalent to 1000 litres (a litre is about a quart).

Sulfur or sulphur?

For historical reasons people in the UK and parts of the British Commonwealth spell the name of element number 16 as 'sulphur', and its derivatives as sulphides, sulphates, etc. The International Union of Pure and Applied Chemists (IUPAC) has deemed the correct spelling should be 'sulfur', and the related words sulfides, sulfates, etc. They are right to do so because the word comes from the Latin *sulfur*. I am torn between the two ways, but I feel that the rulings of IUPAC should be followed.

You might now be wondering why element number 15 is spelt 'phosphorus' and not 'fosforus'. The explanation again lies with its derivation, in this case from two Greek words *phos*, meaning light, and *phorus*, meaning bringing.

*On a time scale 1 ppt is eqivalent to 1 second in 30,000 years.

GALLERY 1

NEARLY AS NATURE INTENDED

An exhibition of some curious molecules in the foods we eat

THERE ARE scores of myths surrounding the things we eat: chocolate is almost addictive; Coca-Cola is just a concoction of chemicals; garlic wards off heart disease and cancer; an aspirin a day keeps the doctor away. None of these statements is true, but they contain a germ of truth. In this gallery we can inspect the portraits of some of the natural and unnatural chemicals which a normal diet contains.

The pleasures of eating are sweet but fleeting, while the warnings about food seem bitter and never-ending. The warnings we should heed are those of professional dietitians, the front-line troops who are fighting the war against poor nutrition and unbalanced diets. While they help the people who are referred to them, the rest of us only hear their advice second-hand, and even then we do not heed it—which may explain why one person in five is now classed as obese (33% or more overweight) in the USA, and one in ten in Britain.

Behind the front-line dietitians is a regiment of armchair food commanders who offer their advice to anyone who listens. Often it is soundly based, telling us how to lose weight and still be properly nourished, but a lot is rather unhelpful, merely condemning some popular foods as 'junk' without explaining why they are so (although this term is generally taken to mean that they contain too much sugar, salt, saturated fats and additives). Examples of junk food are chocolate, colas, hamburgers and french fries. Sadly the healthy alternatives, such as raw celery, mineral water and lentils, lack appeal for many, and especially for children.

Alongside claims about junk food come more dire warnings about the chemicals that are present in other foods, and especially if these have been added merely to make food look and taste more tempting, or if they are there as contaminants that come from pesticides and processing. Surprisingly, most food-related illness comes not from these, but from micro-organisms such as bacteria and fungi, and we are most at risk when we eat food that has not been properly stored or prepared. Ideally food should be free of all dangerous impurities, be they bacteria, fungi or chemicals.

Nature also has its chemicals, and some of these we are rather partial to, such as phenylethylamine and caffeine. Others we try to avoid, such as oxalic and phosphoric acids, and yet others we should take more of, such as salicylate and selenium. In this Gallery we will view a display of molecules that are there in the foods we eat, and all of them are perfectly natural, except one: phthalate (this comes courtesy of the plastics industry). The others are examples of molecules that make us feel better, those which can do us harm, and those which can make us smell.

Three of the molecules are to be found in chocolate. No food provokes the emotional responses of chocolate. To some is it *the* junk food and appears to be little more than a temptation of the devil. Why is it so irresistible? Most people love it, some cannot resist it, and a few unfortunate people have to avoid it. Some people stuff themselves with it until they are sick, while others claim that a mere lick of chocolate will trigger an allergic attack. The makers of chocolate confectionery advertise their products in a variety of ways. They emphasize its wholesomeness and nutritional value, and claim it is full of energy; they suggest you offer it as a gift to a loved one, or even eat it as a way of rewarding and pampering yourself. Whatever its benefits, it has its risks, and some people regard chocolate as junk because its sugar rots teeth, its fats damage the heart, its calories put on weight and its cocoa can trigger a migraine attack.

An analysis of chocolate buyers in the UK showed that most chocolate is bought by women, who account for around 40% of sales, while children buy 35% and men 25%. It tops the list of difficult-to-resist foods, and accounts for over half of all food cravings. Some women even claim to be chocoholics and say they find it impossible to resist, especially before their monthly period. Clearly for them, chocolate is more than just a tasty food or a treat. Chantal Coady, author of the book *Chocolate*, questions whether there really are such people as chocoholics. She writes: 'Although chocolate contains many active chemicals, some of which mimic natural hormones, none of these is addictive.' She believes that women turn to chocolate for consolation when they need a little comfort, and that what they seek is the intense sweetness associated with chocolate confectionery, as well as its luxurious taste and texture in the mouth.

Chocolate is a fairly well-balanced food consisting of 8% protein, 60% carbohydrate, and 30% fat, although this last component is at the upper limit of what is desirable. A normal 100 g (4 ounce) bar provides 520 calories, but it also provides some essential minerals and vitamins:

Minerals		Vitamins	
potassium	420 mg	A	8 mcg
chlorine	270 mg	B1	0.1 mg
phosphorus	240 mg	B2	0.24 mg
calcium	220 mg	B3	1.6 mg
sodium	120 mg	E	0.5 mg
magnesium	55 mg		
iron	1.6 mg		
copper	0.3 mg		
zinc	0.2 mg		

We shall be viewing the dietary importance of minerals in an exhibition in the next Gallery. Looking at this list it is perhaps not surprising that chocolate bars make excellent emergency rations for soldiers and explorers, but there are a few things missing, like vitamins C and D, so it is far from being a complete food. It also has a few other things which are not nutrients, such as phenylethylamine, oxalic acid and caffeine. These have no nutritive value, but they do affect us, and two of them are abundant in other foods and drinks as well. The first three portraits in the exhibition concern these chemicals.

Portrait 1

Aztec dreams—phenylethylamine (PEA)

The only thing in chocolate which comes anywhere near having a feel-good effect on our brain is phenylethylamine (PEA). The Mayas of Middle America, who flourished in Mexico from AD250 to 900, discovered the effects of this when they discovered chocolate, which they took as a drink and which they reserved for the ruling élite. By the time the Spaniards arrived at the end of the fifteenth century, the Aztecs were the dominant civilization and the economy was partly based on cocoa beans—levies from conquered tribes had to be paid in this currency.

Aztec nobles also reserved chocolate for themselves, regarding it as an aphrodisiac, and yet forbidding women to drink it. When cocoa beans were taken back to Europe, chocolate's reputation as a love-stimulant sailed with them. This reputation grew: it was now drunk by both sexes, and in 1624, one author, Joan Roach, devoted a whole book to its condemnation, referring to it with puritanical disapproval as a 'violent inflamer of the passions.' In the eighteenth century the great lover, Casanova, proclaimed chocolate to be his preferred drink.

Cocoa beans are harvested from the cacoa tree, which grows best in warm, moist climates and within 20° latitude of the Equator. The world production of cocoa beans is two million tons a year, and they are grown in Brazil and Mexico for the North American market, and in West Africa for the European market.

After cocoa pods are harvested, the beans are removed and left in the sun to ferment. This exposure turns them brown and converts some of their sugars first to alcohol and then to acetic acid, which we know best in the form of vinegar. The acetic acid kills the shoot and releases other flavour molecules. Phenylethylamine (PEA) forms during this fermentation stage. The beans are then roasted to remove most of the acetic acid, and milled, which causes the cocoa fat to become molten. The extent of the grinding process determines the different grades of chocolate.

Today when we speak of chocolate, we think of a piece of chocolate candy, but originally chocolate was a drink. The name is derived from the Aztec word *xocalatl* meaning bitter water, and it was served as a rather scummy liquid mixed with cinnamon and cornmeal. Later, vanilla and sugar were added to make it sweeter and more palatable for European tastes.

Despite what Casanova thought, chocolate is not an aphrodisiac, but there may be some truth in the idea that it affects the brain. Analysts have detected more than 300 chemicals in chocolate. Two of them are of stimulants: caffeine, which will be dealt with later in this gallery; and theobromine, which is chemically similar and was named after the cocoa tree, whose botanical name *Theobroma cacoa* means 'food of the gods'. Theobromine is also present in tea.

The most likely chemical in chocolate that might explain its feel-good effect is PEA, of which there can be up to 700 mg in a 100 g bar (0.7%). Most chocolate contains much less than this, and a more typical amount would be 50–100 mg. In its pure state PEA is an oily liquid with a fish-like smell, and it can be made in the laboratory from ammonia. (PEA has the curious property of absorbing carbon dioxide from the air.) When people are injected with PEA, the level of glucose in their blood goes up and so does their blood pressure. These effects combine to produce a feeling of well-being and alertness. PEA may trigger the release of dopamine, which is the brain chemical that makes us feel happy, in which case PEA would be acting in the same way as amphetamines such as ecstasy. PEA and ecstasy molecules are roughly the same shape and size, and this has led to the suggestion that they might work in the same way, but scientific proof is lacking that they do.

Our own bodies produce tiny but detectable amounts of PEA naturally, and it is formed from an essential dietary amino acid called phenylalanine. The level of natural PEA varies and it increases when we are under stress. It is also higher than normal in schizophrenics and hyperactive children, but this is more likely to be a symptom of these conditions rather than their cause.

Not everyone can cope with a sudden influx of PEA, which is why some people are sensitive to chocolate, often suffering a violent headache if they eat too much. This happens because the excess PEA constricts the walls of blood vessels in the brain. The human body has little use for PEA and employs an enzyme, monoamine oxidase, to dispose of it. People whose bodies are intolerant of chocolate appear to have difficulty making enough of the enzyme to prevent the PEA building up to levels that triggers migraines.

That PEA is addictive seems unlikely, but there is another reason why some people deny themselves the enjoyment of chocolate. Its fat content, which is called cocoa butter, is primarily a saturated fat. In fact it is 60% saturated, the same as dairy cream, and should be viewed likewise. However, in Dr Hervé Robert's book, *Les vertus thérapeutiques du chocolat*, it is claimed that cocoa butter, unlike cream, does not lead to raised blood cholesterol levels.

The fat in chocolate is rather special in another way. Normal fats are a mixture of saturated and unsaturated fats which tend to soften and melt over a range of temperatures. This is not what we want to happen with a bar of chocolate. Chocolate has literally to melt in the mouth at a temperature of around 35 Celsius, just below the body temperature of 37 Celsius. This is why the best way to enjoy a bar of chocolate is to let a piece of it rest on the tongue until it melts and releases its rich flavour and aroma.

Cocoa butter itself can solidify in several different ways, and each melts at a different temperature. Only one form is right for solid chocolate, which explains why chocolate-making is still regarded as much as an art as a science, and why careful cooling of the molten chocolate is necessary to ensure that the correct form solidifies. If you keep chocolate too long it becomes covered with a greasy white bloom, which makes it look as if it has gone off. It hasn't: this is not a mould, but only another of the crystal forms of cocoa butter, and is perfectly edible.

When chocolate was regarded as a hot drink, the chemistry of its fat hardly mattered. Then in 1847, the Quaker confectioners, J.S. Fry & Sons, of Bristol, England, introduced a solid form that could be eaten as a sweet. They made it by pressing the molten chocolate in order to squeeze out the cocoa butter, and then added this to more molten chocolate. The result was plain chocolate with rather a strong flavour. Much more popular was milk chocolate, bars of which were first produced in 1876 by the Swiss chemist Henri Nestlé. He added condensed milk, which made the product lighter in taste (and colour) and opened the market to children. Other Quaker families—the Cadburys, the Rowntrees and the Hersheys—entered the chocolate business and went on to establish equally large chocolate empires in the UK and the USA.

Since then chocolate has never looked back. Yet it is not without its hidden hazards, although these pose less of a threat than eating too much chocolate, especially if this leads to obesity.

Portrait 2

Rhubarb pie—oxalic acid

Chocolate also harbours oxalic acid, a dangerous chemical that can kill—but rarely does. We take in oxalic acid every day from a variety of sources. It occurs in lots of foods in small amounts, and in a few foods in large amounts, cocoa having one of the largest with 500 mg per 100 g. Green leaf vegetables have the most, such as Swiss chard with 700 mg per 100 g, spinach with 600 mg, and rhubarb with 500 mg. Rhubarb, also known as pieplant in the USA, is popularly thought to have dangerously high levels because this food has killed people. Perhaps less well appreciated is that beetroot (300 mg) and peanuts (150 mg) also have a lot of oxalic acid.

The average person consumes about 150 mg of oxalic acid a day, and in countries where tea is popular the level is generally higher because a cup of tea provides 50 mg. A fatal dose of oxalic acid is around 1500 mg. Could we reach a deadly dose during the course of a normal day? And what effect do even the lower levels of oxalic acid have on us?

Rhubarb is less popular than it used to be, but it was once widely eaten stewed with sugar. It was famed for its laxative properties, and it works because it stimulates our gut to reject the natural toxin, oxalic acid. A bowl of stewed rhubarb could provide us with a sizeable fraction of the toxic dose. To poison yourself by eating bars of milk chocolate would be virtually impossible, no matter how chocoholic you felt, because these contain much less oxalic acid and you would be satiated with them before you reached even the laxative level.

Rhubarb became infamous in World War I when people ate its leaves as a vegetable, and some died through oxalic acid poisoning. The level of oxalic acid in rhubarb leaves is much higher than in the plant stalks, but you are not at risk from eating those.

Rhubarb has long been known as a medicament. In AD 70 the Greek physician and botanist Dioscorides recommended it for treating a variety of conditions. This was European rhubarb, and was used until the twelfth century, when a superior rhubarb appeared from the East. There was much speculation as to where it was grown. Most came from China and it continued to be imported on a large scale as powdered root for hundreds of years.

Rhubarb root has been used in traditional Chinese medicine for more than four thousand years. The Royal Society of Arts, Manufacturers and Commerce decided to promote the cultivation of the new rhubarb in the British Empire, and in the 18th and 19th centuries it awarded several gold medals to those who grew the best varieties. In 1784 the Swedish apothecary Carl Wilhelm Scheele detected oxalic acid (which he knew as acid of sorrel) in rhubarb roots, and showed that the amounts in the plant's leaves were too large for these to be edible. The oxalic acid is thought to be the plant's protection against cattle. In 1860 the Victorian best-seller, *Mrs Beeton's Book of Household Management*, reported that rhubarb was in every kitchen garden, and she gave recipes for rhubarb pies, rhubarb jam and even rhubarb wine. The easiest way of cooking it was to stew the chopped stalks with sugar. When aluminium pans became popular, stewing rhubarb was discovered to have another bonus: it cleaned the pans beautifully. It did this because the oxalic acid dissolved off the top layer of metal, although the amount that was removed this way was so tiny it was no threat to health.

This affinity of oxalic acid for metals also explains another curious anomaly, and is the reason why nutritionists refer to it as an anti-nutrient. Oxalic acid interferes with the essential minerals iron, magnesium and especially calcium. Earlier this century spinach was advocated as a rich source of iron, and indeed it has higher levels of this metal than most vegetables. For example spinach has 4 mg of iron per 100 g, compared to peas which have 2 mg, brussels sprouts 1 mg, and cabbage only 0.5 mg. Despite having more iron, the oxalic acid in spinach renders 95% of this metal useless as a nutrient, and only 5% can be absorbed by our body. The cartoon character Popeye attributed his strength to spinach, but he was sadly misinformed. By all means eat spinach as a vegetable, enjoy it, but expect very little from it except a modest amount of vegetable protein and a little vitamin C.

Oxalic acid kills by lowering the calcium in our blood below a critical level. (The antidote is calcium gluconate.) Calcium is essential for keeping the blood at a constant level of acidity and viscosity, as well as for clotting and transporting phosphate around the body. But even in non-lethal doses the effect of oxalic acid on calcium is worrying, because it forms insoluble calcium oxalate, crystals of which can grow into painful stones in the bladder and kidneys. The development of these is more likely if our fluid intake is low. Doctors whose patients are prone to develop such stones put them on a low-oxalic acid diet, which excludes the foods we have been talking about. Although such foods can be avoided, we cannot exclude oxalic acid entirely from the body because there are other sources. For example, surplus vitamin C, which the body cannot store, may be turned into oxalic acid, and a side-effect of taking massive doses of this vitamin may also be kidney stones.

According to John Timbrell, a toxicologist at the London School of Pharmacy and author of *Introduction to Toxicology*, it is possible to get a fatal dose of oxalic acid in other ways. People who have accidentally or deliberately drunk ethylene glycol, which is used as antifreeze in cars, may die of oxalic acid poisoning because the body converts ethylene glycol to oxalic acid.

Plant cells are known to make use of oxalic acid, but it has no role in animal cells—or so it was once assumed. Recent research suggests otherwise. Despite the apparent toxicity of oxalic acid, the body will tolerate surprisingly high levels of it, and research scientists in Germany discovered that human tissue contains more oxalic acid than was previously suspected. Dr Steffen Albrecht and co-workers at Dresden University have challenged the view that oxalic acid is merely an unwanted end-product of metabolism. They have developed a sensitive method of analysis for oxalic acid, and can measure concentrations as low as millionths of a gram (mcgs) per litre of blood. Their work has revealed markedly different levels of oxalic acid in the blood: plasma, the fluid part of the blood, has 400 mcg per litre while serum, the clear solution which separates from blood after the plasma has coagulated, has 1200 mcg. Some blood cells have 250 000 mcg, which sounds a lot but when converted to milligrams is only 25 mg per 100 g, which is low compared to the levels in some foods. Albrecht says the high levels of oxalic acid point to its having an active role in human metabolism, although what that is remains unknown.

Oxalic acid is made commercially by treating either sugar with nitric acid or cellulose with sodium hydroxide. The acid is very soluble in water—a litre will dissolve 150 g—and it forms a corrosive solution. Industrially it is used in tanning leather, dyeing cloth, cleaning metals and for purifying oils and fats. The only guise in which it might be found in the home is in stain-removers for iron-based stains, such as rust and ink from fountain-pens.

Portrait 3

The Coca-Cola conundrum—caffeine

Chocolate contains a little caffeine, but there are much richer sources, such as coffee, tea and cola. Most ingredients in colas have been criticized at one time or another, and yet the young people of the world continue to love them. But look at the label on a bottle or can and it appears that all you are drinking is a solution of chemicals in fizzy water. Colas contain little that can be described as natural. The main ingredients are sugar or an artificial sweetener, phosphoric acid, caffeine, and a blend of flavourings, which are supposed to be a secret. Once upon a time they were, and when Coca-Cola was first invented this was part of its attraction.

There is no denying that the secret formula of Coca-Cola has been highly successful: it has seduced the taste buds of billions of people around the world. Nor should we be surprised, because it is a refreshing, pleasantly flavoured drink, and a can of ice-cold coke can really quench your thirst on a hot summer's day. Not surprisingly, it has many imitators.

The story of Coca-Cola began on 28 June 1887 in Atlanta, Georgia, when a pharmacist, Dr John Pemberton, then 56 years old, was granted the trademark, Coca-Cola, for a drink he had invented. The time was right for the new beverage because the city of Atlanta had just voted to ban alcohol, so perhaps it was not surprising that Coca-Cola sold well. Pemberton's new drink continued to sell even after prohibition was repealed in the city later that year.

Pemberton placed an advertisement in the *Atlanta Journal* describing his new drink as follows:

Delicious! Refreshing! Exhilarating! Invigorating! The new and popular soda fountain drink contains the properties of the wonderful coca plant and the famous cola nut.

And indeed the drink is still named after these ingredients—the coca plant, which is the source of cocaine, and the cola nut, which is rich in caffeine. I should hasten to add that neither of these plants provides ingredients for today's colas.

Pemberton had stumbled upon a recipe that was to become the world's best-selling soft drink. He kept the flavour ingredients a closely guarded secret, and the Coca-Cola company claim that only the top two executives in the company know what they are, and how they should be blended.

Most of the main ingredients of Coca-Cola have always been common knowledge: sugar, caramel, caffeine, phosphoric acid, lime juice, and vanilla essence. Together these make a passable concoction and the caramel, lime and vanilla are the dominant flavours. There never was any cocaine in Coca-Cola, although Pemberton, who was a regular user of this drug, may have experimented with it. Cocaine was certainly added to certain tonic wines of the day: Queen Victoria herself was reputed to be very partial to some of these. The cola extract was also dropped from the recipe early on, in preference to adding purified caffeine directly. The level of acidity of Coca-Cola, which was needed for its refreshing taste, was originally due to citric acid, which occurs in citrus fruits, but this ingredient was soon replaced with cheaper phosphoric acid.

Pemberton needed to make his new drink distinctive, and so he experimented with other flavours in smaller amounts. Finally he found a blend that he liked and gave it the code name 7X. So carefully did the Coca-Cola company protect the secret of 7X over the years that it was prepared to defy court orders rather than reveal it. In India manufacturers are obliged by law to say what is in a drink, but in 1977 the company decided to cease marketing Coca-Cola in India rather than reveal its secret.

Over the years there have been several attempts to guess what 7X consists of. Its natural essences are present in only tiny amounts, so it was almost impossible to discover what these were by chemical analysis, because each essence consisted of numerous flavour chemicals. In 1983, William Poundstone, author of the book *Big Secrets*, printed his list of what he thought was in 7X, which he said consisted of orange, lemon, nutmeg, cassia, coriander, neroli and lime. Cassia is also known as Chinese cinnamon, and neroli is extracted from the bitter-orange flower. Poundstone had made quite a shrewd guess, as we shall see.

Modern analytical techniques will lay bare the intimate details of any secret mixture, and so perhaps it was not unduly worrying for Coca-Cola when Mark Prendergast finally published the recipe for 7X in his book *For God, Country and Coca-Cola* in 1993. He says he came across it in the tattered remains of one of Pemberton's laboratory notebooks in the company archives. The mysterious 7X was a blend of the oils of lemon (120 parts), orange (80), nutmeg (40), cinnamon (40), neroli (40) and coriander (20). Pemberton mixed these with alcohol and then left it to stand for 24 hours to give him his secret extract. Today there is no alcohol, but it may be that the use of that ingredient during prohibition explains Pemberton's original need for secrecy.

It is still claimed by the Coca-Cola company that it is the sequence in which the ingredients of 7X are blended that is the key to producing the 'real thing', and it may still be that only two executives of the company know this. There are people who say they can identify the different colas that are available, but a discriminating palate for this type of drink is never likely to be considered the mark of a connoisseur. Colas are simply refreshing drinks that do no harm and keep many people employed in making, distributing, selling and advertising them. When you buy a can of cola, it hardly matters that packaging, promotion and profits account for 95% of what you spend. You can be under no illusion that you are buying something essential, and a glass of water is even better at quenching your thirst and costs virtually nothing. What you are really buying is a solution of caffeine, and this can have an effect on you.

The amount of caffeine in a can of cola is 40 mg, the same as in a cup of tea, and about half that in a cup of freshly ground coffee. The same volume of instant coffee provides 60 mg, and over the last 50 years this has become the most popular way to take caffeine. Instant coffee was first produced by the Swiss company Nestlé in 1938, and sold as Nescafé (the Brazilian Institute for Coffee had shown that coffee could be reduced to a soluble powder in 1930). Instant coffee really came into its own in World War II when it was widely used by US troops, and thereafter it became part of everyday living.

Young people may get their daily dose of caffeine from colas, but most adults get it from coffee and tea. While flavour is the most important part of these drinks, it is the caffeine that explains their enduring popularity. Tea is mainly drunk in the countries in which it is grown, such as India, Sri Lanka, and especially China, but a few countries are large importers, such as Great Britain and Australia. Coffee, on the other hand, is mainly grown as a crop for export in countries like Brazil, Colombia, Indonesia and Kenya. International trade in coffee beans exceeds $7 billion a year, making them one of the top four traded commodities (along with coal, grain and oil).

Worldwide consumption of caffeine is now estimated to be over 120 000 tons per year, which works out at about 60 mg per person per day. Scandinavians have the largest caffeine intake, generally from coffee, with over 400 mg per day; the British consume around 300 mg per day, much of it as tea; and the Americans, long regarded as big coffee and cola drinkers, get a surprisingly low 200 mg per day.

A fatal dose of caffeine taken by mouth would be about 5000 mg, the amount in 80 cups of coffee or 120 cups of tea. When you take in caffeine your body mobilizes its defences to dispose of the invading toxin, which is how it sees this non-nutrient. It rids itself of the offending molecule by plucking off carbon atoms, although at first this has little effect, because it leads to new molecules, such as theophylline and paraxanthine, which are just as potent. However, the process continues and finally the product is xanthine, which the body can eliminate in the urine or put to other uses. All this explains why the effects of caffeine in the body persist for around five hours. Curiously, cigarette smoking stimulates the liver to generate more caffeine-destroying enzymes and for smokers the effects last about three hours.

More than 60 plant species produce caffeine, and it is believed that this chemical probably protects them against attack by insects. The coffee bush is indigenous to Ethiopia and was cultivated there over a thousand years ago. It reached Europe around AD 1600, probably via Turkey where it got its name, kehveh. Tea has a much longer tradition, and was being drunk in China in 2500BC, but it too did not reach Europe until the seventeenth century. The cola, or kola, plant is an evergreen tree of tropical Africa which produces glossy nuts with a high caffeine content. The way to release their caffeine was to chew them.

Caffeine is not only a pick-me-up: it also has medicinal benefits, and is used in painkillers, asthma treatments, and diet aids. These rely on its effect of stimulating the metabolism and relaxing the bronchial nerves. Caffeine has long been used to increase physical endurance. In Tibet, not only do Tibetans drink a lot of tea themselves, but they give their horses and mules large vessels of the drink. Distances were once measured in the number of cups of tea deemed necessary for a journey: three cups of tea would give you enough 'fuel' for around 8 kilometres.

Chemically, caffeine is a white powder which was first isolated in 1820 by the German chemist Friedlieb Ferdinand Runge, but it was not until 1897 that its molecular structure was deduced. It can be made in the laboratory, but the commercial market is supplied by the caffeine which is produced as a by-product of decaffeinating coffee. Removing caffeine without affecting the taste of coffee is relatively simple, and involves extracting it with liquefied carbon dioxide.

There are many popular myths about caffeine. It is accused of causing sleepless nights, indigestion and bad breath, and as if that were not enough it has also been blamed for raising cholesterol levels in those who drink a lot of it, and so putting them at risk from heart disease. There was even a suggestion in the 1970s that it might cause liver cancer, but that scare turned out to be completely unfounded. Nor does it cause insomnia, indigestion or heart disease, and this was the conclusion of 175 scientists from around the world who attended the International Caffeine Workshop which was held in Greece in 1993. As more and more data have been collected and analysed, the many scares about caffeine have been shown to be little more than the artefacts of poorly designed epidemiological studies into people's eating habits.

Caffeine affects us in many ways. It is metabolised by the liver, which takes about 12 hours to remove 90% of any caffeine we have consumed. The first few times we have caffeine it raises our heart rate and blood pressure quite dramatically, but as we become regular drinkers of colas, coffee and tea, our body stops reacting in this way. Because of these physiological effects it was not surprising that caffeine was thought to be a factor in some common diseases. A report in 1973 suggested that the risk of thrombosis was doubled if a person consumed 400 mg a day, equivalent to drinking five cups of fresh coffee. However, a study in 1990 on 45 000 men failed to find any connection between thrombosis and coffee drinking. A supposed link with heart disease was also shown to be wrong by a large survey in Scotland, where both men and women have a particularly high incidence of this condition. The researchers there questioned more than 10 000 middle-aged men and women and could find no link between caffeine intake and heart disease.

Caffeine acts as a stimulant and its drinks are advertised to emphasize this, so we are told that coffee wakes us up and colas refresh, while a cup of tea revives. It works by boosting the brain's own stimulant dopamine, and this responds up to around four cups of coffee, after which extra cups have no more effect on the level of this in the brain. Caffeine in excess is popularly believed to keep us awake at night, but it probably does not have this effect on most people, unless they drink too much. There are those who metabolise caffeine only slowly, and they may suffer this way. Despite earlier reports that we cannot become addicted to caffeine, caffeine withdrawal symptoms now seem to be accepted, and they are, in order of occurrence: headache, depression, fatigue, irritableness, nausea, vomiting.

In addition to its caffeine, tea may have hidden benefits in the form of three other chemicals it contains. These are salicylate, epicatechin gallate and epigallocatechin gallate. We shall look at the portrait of salicylate a little further along in this gallery. The other two molecules are part of a group known as flavinoids, and are thought to protect the body against free radicals. These are highly dangerous natural chemicals which have a rogue electron, and it is this which enables them to attack key components of the living cell such as DNA, thereby possibly causing cancer. It is their relentless attack on the body which is thought to be the underlying cause of ageing.

Perhaps tea-drinking can help in the fight against free radicals. A Dutch research team carried out a 15-year study on men aged 50 and over, and in 1996 reported their findings which showed that tea-drinkers had a much reduced incidence of stroke. They attributed this to the flavinoids capability of destroying free radicals. Other research has shown that the tea flavinoids also protect against tumours, at least in animals.

Portrait 4

Rust remover—phosphoric acid

The ingredient of colas which looks rather odd, and rather menacing, is phosphoric acid. Generally we are more familiar with this acid as the active agent in rust-removers, and with its salts, which are called phosphates, and are used in detergents. In the 1970s and 80s, phosphates became a dirty word, and were blamed for the pollution of rivers and lakes, with detergents being fingered as the most important source. We will look a little closer at this issue in Gallery 6, where there is a portrait of phosphates.

People need phosphate in their diet as an essential nutrient to make DNA, build their bones and form their membranes. It also is needed for the chemical adenosine triphosphate (ATP), which plays a central role in helping get the energy we need from food. Phosphate-containing molecules also act as messengers, and govern calcium transport. In addition to these major roles phosphate has many minor uses in the body. It might seem that such a key element could be in danger of being in short supply in our diet, but this rarely happens because the body recycles it very effectively and in any case we have an enormous store of phosphate in our skeleton. The phosphoric acid in colas can be regarded as making a useful but minor contribution.

Sometimes the phosphoric acid in colas has found other uses. Motorists, motorcyclists and truck drivers in the 1950s and 60s used cola to clean the chrome bumpers (fenders), grilles and headlights which lavishly adorned their vehicles in the fashion of those times. The phosphoric acid reacted chemically with the chrome to form a hard surface layer of chromium phosphate which protected it. It also dissolved any rust that formed and protected any of the underlying steel which had become exposed. Industrially phosphoric acid is still used for this purpose, and all anti-rust paints rely on it.

There is nothing sinister about phosphoric acid or its salts. To say that colas contain an industrial cleaner, as one book has claimed, is strictly true, but this is no reason to not drink them. Any phosphate in our food becomes phosphoric acid in the acidic conditions of the stomach. Every living cell needs phosphoric acid to function and it matters not where it comes from.

The phosphoric acid in colas presents no threat to health; indeed, we could regard it as an essential mineral. Plants begin the process of supplying the food chain with phosphate by extracting it from the soil, and they store phosphate in their seeds as the chemical phytic acid. They use this store when they germinate so they can put down roots without needing to take in any phosphate from the environment. Although seeds are highly nutritious on account of the protein, carbohydrate, fats and minerals they contain, they provide us with little in the way of phosphate because we cannot digest the phytic acid store since we lack the enzymes to release its phosphoric acids. So the phytic acid passes straight through us—not that we need it, because every plant and animal cell that we eat contains more than enough phosphate.

We get most of the phosphate in our diet from natural sources such as fish, meat, eggs and dairy products, and a little from unnatural sources such as colas, processed cheese, cheese spreads, sausages and cooked meats, to which it is added to improve texture and regulate acidity.

Portrait 5

The curse of the cure-all—dipropenyl disulfide

When is something simply a flavouring and when is it a medicine? Garlic is admired by many for having both these properties, and the chemical which is responsible is a simple molecule called dipropenyl disulfide. But can it really be a healing drug? And if it is, should it not be subject to the same kinds of tests that all pharmaceutical drugs are exposed to, so that we know it works and that it is perfectly safe?

Of course you don't need to bother with such tests if the material you are testing is basically a food flavouring ingredient that has long been part of the human diet. Time has tested it for you, although sometimes Time can be proved wrong—witness the once popular herb, comfrey, long used in salads and to make comfrey tea. The sale of comfrey is now banned in Europe because of the harmful chemicals it contains. So perhaps chemists are not being too finicky when they suggest that everything that purports to be a healing drug should be tested in the same way as pharmaceuticals. In other words we should put dipropenyl sulfide through a programme of tests on animals such as mice and rats, dogs and cats, and finally monkeys and humans. Of course it would fail early on, because it has undesirable side-effects, the worst of which is to give the patient an advanced case of halitosis. No creature deserves to have this obnoxious material forced on them, except human volunteers. Nevertheless, it is perfectly natural, and it is the most popular of the so-called alternative medicines on sale today. It is sold as garlic oil and is purchased by millions of people all over the world and taken in the form of capsules. In Germany it is the best-selling over-the-counter drug. We can get used to garlic and eventually come to like it.

Garlic-growing is big business, as well as being an essential part of the domestic garden in many countries. The USA produces around 65 000 tons a year, worth $180 million, and this is grown mainly in California, especially around the small town of Gilroy (pop. 33 500), at whose annual garlic festival it is possible to consume garlic ice cream, garlic cheesecake and garlic scones. In Europe garlic tends to be used in cooking, especially for casseroles and soups, or in salads, and all over the world garlic bread has become a popular way of enjoying it.

Garlic used in cookery loses much of its sting while adding piquancy to soups and savoury dishes. Uncooked garlic in salads can be enjoyable to the eater but not to those they come in contact with afterwards. Yet some people prefer to eat it raw for health reasons, believing it is effective in warding off cancer and cardiac disease. Those who eat it regularly may find their bad breath protects them against illness because it keeps others at a distance. Many are even willing to take it daily in large amounts as though it really was a medicinal drug—but it isn't.

The active ingredient in garlic, dipropenyl disulfide, has two sulfur atoms at its centre, and it is these which produce the odour that its users have to endure, along with their family and close friends. Any chemical we consume which has a lot of sulfur, such as garlic, onions and certain forms of protein, poses a slight social problem for us, if not for our body. One way to get rid of some of the sulfur is to turn it into the obnoxiously smelling molecule methyl mercaptan, which we can breathe out. This is the main cause of halitosis, and we shall look more closely at its portrait next.

A clove of garlic has almost no smell until it is cut or crushed, but when this happens an enzyme called allinase works on an amino acid called allin, and converts it to allicin which is the main constituent in garlic extract. This is the precursor of dipropenyl disulfide—basically the same molecule but with an oxygen atom bonded to one of the two sulfurs. Allicin easily loses its oxygen atom and reduces to the more volatile molecule dipropenyl disulfide. This is the compound which gives garlic its odour.

Raw garlic will give you plenty of this disulfide, but cooking gets rid of it because it is volatile enough to evaporate during cooking. This is the reason you can safely eat a soup or stew that has lots of garlic in the recipe, and still enjoy a friendly tête-à-tête with someone. Some claim that eating parsley or lettuce with raw garlic actually neutralizes its odour, which may well be so, but the evidence is not compelling, and in the end some of it will still be exhaled on the breath.

Epidemiological studies in China and Italy reported that garlic eaters had fewer gastric cancers, and a survey of 40 000 US women appeared to show a link between garlic consumption and lower rates of colon cancer. However, when Elisabeth Dorant and colleagues at the University of Limburg, Maastricht, fed laboratory animals fresh garlic or garlic extracts, they did not observe fewer cases of cancer, although they found tumours were slightly slower in growing in those rats with the disease.

Garlic is known to lower blood cholesterol levels by 10%, if you eat a clove a day, and so it might help prevent cardiovascular disease. However, the evidence that it does so is again less than convincing. In 1994 Christopher Silagy of Flinders University, Adelaide, and Andrew Neil of Oxford University reviewed several tests of the effect of garlic on blood pressure. They concluded that it only helped those with slightly elevated blood pressure, and they could not recommend it as routine clinical therapy.

Such scientific evidence will not impress those who are still convinced that garlic harbours something rather remarkable, and they can point out that garlic has been used in medicine for hundreds of years. Garlic's supporters can even validate their cure-all claims with a bit of chemistry by pointing out that allicin and dipropenyl disulfide are antioxidants, and as such are highly regarded because they can mop up peroxides in the body, thereby preventing the formation of free radicals.

Whether garlic really is effective in warding off cancer and heart disease is doubtful, but the plant is not without its uses. Indeed, it is essential at Halloween, when ghosts walk, witches dance, demons pounce and vampires feast. This is the time to get out the garlic, which is guaranteed to be 100% effective in warding off evil spirits. What probably deters them is the smell of methyl mercaptan on their victim's breath, and this molecule is the subject of our next portrait.

Portrait 6

The worst smell in the world—methyl mercaptan

There are official standards for acceptable levels of noxious smells, and methyl mercaptan heads the list. This molecule makes the news wherever it is emitted in large amounts, and sometimes it does so because it is used industrially, for example for making the insecticide Dimethoate. When it was accidentally released from a factory in Waltham Abbey, England, the local residents were so sickened by its smell that some rushed to hospital assuming they were being poisoned by a deadly pollutant. Others rang the local gas company. This is not as surprising as it seems, because compounds similar to methyl mercaptan are used to odourize natural gas, so that leaks are easy to detect.

Methyl mercaptan is also produced naturally from bacteria in the environment, and the shoreline near Edinburgh, Scotland, often exudes it, much to the distress of the residents of the select suburb which overlooks that beautiful stretch of water.

The methyl mercaptan we breathe out after eating garlic or taking a garlic capsule is produced in the body as we digest allicin. Bacteria are also responsible for the methyl mercaptan we generate in our own mouths and may breathe out continuously as bad breath. This is formed from our own body protein as it breaks down under bacterial attack. We can easily detect methyl mercaptan when someone speaks to us—humans can detect it at levels of parts per billion in air—but, curiously, we cannot smell the gas we produce ourselves. In Japan, you can test your own breath with an Oral Checker of which thousands have already been sold. Katunori Nakamura has patented his halitosis detector, which is about the size of a powder compact and works on the principle that a metal oxide, like tin oxide, changes its electrical resistance when it absorbs a gas like methyl mercaptan onto its surface.

Bad breath is caused by several molecules, such as hydrogen sulfide and dimethyl sulfide, but the main culprit is methyl mercaptan. Hydrogen sulfide, the traditional stink of the chemistry laboratory, is much less smelly, and the same is true of dimethyl sulfide, which is part of the aroma of fresh coffee. Graham Embery of the University of Wales at Cardiff researches the sulfur-containing molecules found in the mouth which arise as a result of the activity of bacteria. These break down the protein residues and release methyl mercaptan from the amino acids cysteine and methionine. If the smell of methyl mercaptan is very strong, it indicates gum disease. Methionine is essential for all living things, and animal protein contains up to 4% of this amino acid; therefore bacteria are capable of releasing enough methyl mercaptan to make the victim's breath smell vile.

Embery and Gunnar Rolla of the University of Norway in Oslo are authors of the book *Clinical and Biological Aspects of Dentifrices*, which devotes a whole chapter to halitosis. Embery's advice to those who suspect that they exhale methyl mercaptan is to use a toothpaste that contains anti-plaque agents, such as zinc or tin salts. These metals interfere with the enzymes in the bacteria which produce the methyl mercaptan. Traditionally, mouthwashes are supposed to cure halitosis, but they do little more than clean the mouth and disguise the offending smell. The best known one, Listerine, consists of water and alcohol, with benzoic acid and natural flavours, such as thymol and menthol. A good rinse with a mouthwash will remove about half the oral bacteria. A more popular way to clean the mouth is to increase saliva flow by chewing gum.

Our feet can also harbour microbes that give off methyl mercaptan, especially if we provide them with a perfect environment of unwashed socks and unventilated shoes. *Staphylococci* and *aerobic coryneform* bacteria are to blame, and these flourish in the increasingly alkaline conditions which are a feature of such socks and shoes. If you have smelly feet, then the chemical answer is to insert into your shoes special charcoal-filled insoles, which have layers of carbon that absorb the methyl mercaptan. Since the amount of methyl mercaptan is tiny, the insoles will go on working for weeks.

Methyl mercaptan is the simplest member of a series of compounds in which there can be chains of up to 20 carbon atoms attached to a sulfur atom. Methyl mercaptan has one carbon atom. Mercaptans with three and four carbons are those we encounter when we smell a leak of gas. A mercaptan with 18 carbons attached to a chain is used as a wax in silver polishes.

A major drawback in manufacturing and transporting methyl mercaptan for industry is its low boiling point of 6 Celsius. Luckily, it can easily be turned into the chemically similar dimethyl disulfide (DMDS), a yellow liquid which boils at 110 Celsius. This consists of two methyl mercaptans joined through their sulfur atoms. It is only slightly less odorous, but is much safer to transport, and most is made at Lacq in south-west France, where natural gas wells bring up large amounts of hydrogen sulfide. This is reacted with methanol to form methyl mercaptan and then converted to DMDS.

Methyl mercaptan is used industrially to make pesticides, and especially for weedkillers for cereal crops like wheat, maize and rice. Its chief use in industry is to regenerate the catalysts used in the refining of petrol. Methyl mercaptan is also used to make methionine, an amino acid which may be deficient in the diet. Some animal feeds are now fortified with methionine, thereby increasing the amount of this in the animals' meat and milk.

Portrait 7

Chinese medicine—selenium

Methyl mercaptan and dimethyl sulfide may be the worst smells we come across in normal life, but there are even worse variants of these molecules: their selenium versions. Selenium is chemically very similar to sulfur, but when it replaces sulfur in a volatile molecule the smell intensifies dramatically. Research chemists who work with selenium compounds have to be very careful to avoid contact with them. Any that gets on to the skin, or even on to clothing, is liable to be expelled as a methyl compound by any micro-organism that is around. If you accidentally ingest some, then your breath will smell appalling. If you take too much selenium then you could even poison yourself.

Despite this unpleasantness, selenium is essential to many species, including humans. We need it only in microgram quantities, but even so every cell of our body contains over a million atoms of selenium. At such low levels it poses no threat to our social life.

It is difficult to measure how much selenium we take in, how much we excrete, and how much we really need. The daily intake varies between 6 and 200 mcg according to the type of food we eat. The average Westerner takes in about 60 mcg per day, which is more than enough to prevent the symptoms of selenium deficiency—a mere 10 mcg may be all we need, provided we get it regularly. Some days our body may lose more selenium than it absorbs, but because the average adult holds about 15 000 mcg (or 15 mg) this does not pose a threat. A single dose of 5000 mcg (5 mg) would be dangerous, and 50 000 mcg (50 mg) would be lethal for many humans. We store most of our selenium in our skeleton, but the parts of the body with the highest levels of selenium are hair, kidneys and testicles.

Most people get most of their selenium from wheat products such as breakfast cereals and bread. The foods richest in selenium are:

- seafoods, such as tuna, cod and salmon
- offal, such as liver and kidney
- nuts, such as Brazil nuts, cashews and peanuts
- wheat germ, bran and Brewer's yeast.

All these have 30 mcg or more of selenium per 100 g of product, although in the case of wheat and meat products the level depends upon the soil of the farm from which they came. The only people who might just be at risk of selenium deficiency are pregnant and breast-feeding women, and children—and only if they carefully avoided all the types of foods listed above. Essential though selenium is, we can have too much, and the recommended maximum daily intake is 450 mcg. Above this we risk selenium poisoning, the most obvious symptom of which is bad breath and body odour. The smell is caused by volatile methyl selenium molecules which our body produces as it rids itself of the selenium it does not need.

Despite the smell, we would die without selenium. In 1975 it was proved essential for humans when Yogesh Awasthi discovered it was part of a human enzyme called glutathione peroxidase. In 1991 Dietrich Behne in Berlin found selenium in a second enzyme, deiodinase, which promotes hormone production in the thyroid gland. If the amount of selenium in our body is too low, then we appear to be at risk from several conditions, such as anaemia, high blood pressure, infertility, cancer, arthritis, premature ageing, muscular dystrophy and multiple sclerosis. As yet there is no proof that a lack of selenium in the diet leads to these conditions, and it is more likely that this element is having a second-order effect, in other words it controls other components which actually do the harm.

Selenium is known to protect us against the effects of other toxic metals such as mercury, cadmium, arsenic and lead: for example, the damage that cadmium can do to our reproductive organs and to a foetus is thought to be prevented by selenium. Tuna fish, which accumulate higher-than-expected levels of mercury, are also thought to be protected by selenium, and analysis shows that for every mercury atom in a tuna there is also an atom of selenium. This 1:1 ratio appears to be true for other marine mammals such as seals, and for men who work in mercury mines.

Most normal diets contain more than enough selenium, so there is little need for people to take selenium supplements, although these are on sale in health foods shops and pharmacies. As a dietary supplement, selenium is taken in the form of sodium selenite, which is a white crystalline material that is soluble in water. The daily dose is 50 mcg. Selenium was first popularized as a dietary aid by Alan Lewis, whose book *Selenium: the Essential Trace Element You Might Not be Getting Enough Of* was published in 1982. Lewis reported that selenium could be used to treat rheumatism, arthritis, heart disease and cancer, and that it would even delay old age. While most of these claims appear fanciful, and are based mainly on anecdotal evidence, two at least are well founded: trials in China have shown that selenium does prevent certain types of heart condition, and that the body needs a certain level if it is to ward off cancer effectively.

The Chinese have long had a special interest in selenium because large areas of that country have selenium-deficient soil, and this affects the health of the local population. Children in the Keshan region were prone to a heart condition known as Keshan disease, which is caused by a lack of selenium. This disease results in a swelling of the heart and kills half of those afflicted. A large-scale trial in south China in 1974 involved 20 000 children, half of whom were given selenium-containing tablets and half given a placebo. Of those on the placebo, 106 developed Keshan disease and 53 died, while of those on the selenium supplement only 17 got Keshan and one died.

Another test in China also found that selenium was beneficial in reducing cancer cases. Among the Chinese people living in the north central province of Linxian there is a high incidence of stomach cancer. The people there agreed to take part in a five-year project and 30 000 middle-aged people were given different combinations of dietary supplements such as vitamins A, B_2, C and E, zinc and selenium. The study showed a remarkable drop in cancer cases in the group taking vitamin E plus selenium.

Selenium was discovered in 1817 by Jöns Jacob Berzelius at Stockholm, Sweden. He named it after *selene*, the Greek word for the

Moon, to match the name of the related element tellurium, which was based on the Latin *tellus* meaning Earth. He found it when he investigated a red-brown sediment which collected at the bottom of the chambers in which sulfuric acid was made. The element selenium is available either as a silvery metal or as a red powder. The main producing countries are Canada, the USA, Bolivia and Russia, and most comes from copper smelters and refiners. Copper sulfide ores have copper selenide as an impurity. The most important source of selenium is the slime which settles at the bottom of tanks when impure copper is refined electrolytically, and this sediment may contain up to 5% selenium. This source accounts for about 90% of selenium production. Each year about 150 tons of selenium is recycled from industrial waste and reclaimed from old photocopying machines.

The metallic form of selenium has the curious property of generating an electric current when light falls on its surface, and it is used in photoelectric cells, light meters, solar cells and photocopiers. These electronic uses account for about a third of all selenium production, and require high grade selenium of 99.99% purity. The second largest user is the glass industry, where selenium goes into special glass such as the bronze architectural glass which screens out the Sun's rays. The third main use is to make sodium selenite for animal feeds and food supplements. Selenium is also used in metal alloys, such as the lead plates used in storage batteries; in rectifiers to convert electric current from a.c. to d.c.; and in anti-dandruff shampoos.

Selenium is rarer than silver, and one day mineral sources of the element will be exhausted. Then we may have to harvest it by growing crops like milk-vetch on high-selenium soils. This could yield as much as 7 kg per hectare (3 kg per acre). The current world demand for selenium, of about 1500 tons per year, would require about 200 000 hectares to be farmed this way. But as reserves of selenium in ore deposits amount to over 100 000 tons, it will be quite some time before this type of farming will be needed.

The effects of high-selenium soils have been known for a long time. Animals grazing on such pasture may suffer from the so-called 'blind staggers'. Marco Polo (1254–1324) wrote that the animals of Turkestan behaved this way. The plant responsible for the staggers was probably vetch, which can concentrate up to 1.4% of its weight as selenium. The cowboys of the Wild West knew that this plant could affect their herds, and called it locoweed, from the Spanish word *loco* meaning insane. In 1934 the biochemist Orville Beath proved that the staggers were caused by excess selenium in the diet. When the vetch had an offensive smell, it was a sure indication that it had absorbed a high level of selenium.

Portrait 8

The state of the heart—salicylates

In 1763 the Reverend Edmund Stone, an English parson living in the Cotswolds, made an infusion of the bark of the white willow and gave the drink to people in his village who had fever. Today we can only guess at what his parishioners were suffering from, but one suspects most of them probably had a mild virus infection, such as 'flu. In any event, they clearly had high temperatures and the treatment was successful in bringing these down. We now know that it would have been effective because Stone had given the villagers a solution that in the human body would produce salicylic acid, which is good at reducing high body temperatures.

In the century which followed, this simple but effective treatment continued even though it had unpleasant side effects. Salicylic acid is a strong irritant, causing bleeding and ulcers in the mouth and stomach. It was not until two chemists working for the German chemical company Bayer made the derivative, acetylsalicylic acid, that the treatment became relatively safe. That was in 1893, and the chemists were Felix Hoffmann and Heinrich Dreser. Their product was named aspirin, and for over a century it has brought relief to millions of people around the world. Aspirin works by blocking an enzyme that makes prostaglandins, the chemicals which signal that the body has been injured or invaded by a micro-organism. Prostaglandins are generated in excess, and the result is inflammation, pain and fever.

Today in the USA around 20 *billion* aspirin tablets are taken each year, even though it is still a risky remedy and can cause stomach inflammation in some people. The best known form of aspirin is Alka Seltzer, whose tablets also contain citric acid and sodium bicarbonate. The bicarbonate reacts with the aspirin to form its sodium salt, thereby making it soluble in water and supposedly quicker acting, and reacts with the citric acid to generate bubbles of carbon dioxide. The citric acid also masks the taste of the aspirin.

Although aspirin has been used for a long time it is not without its more serious risks, and for some young children aspirin has proved fatal, when they have been given it to treat a viral infection like 'flu or chicken pox. They developed what is known as Reye's syndrome, and although this is extremely rare, it is best never to give aspirin to a child under the age of 12.

Despite its disadvantages, aspirin is much more than just a painkiller, and is prescribed by doctors to patients who have suffered a heart attack because it inhibits the formation of those chemicals which cause blood platelets to aggregate together, which is what starts a blood clot. Aspirin is normally sold as 300 mg tablets, and these can safely be taken at a rate of two every four hours to a maximum dose of a dozen tablets (4 g) per day. A single dose of 10 g (30 tablets) can kill an adult because it makes their blood too acidic. The body tries to cope by rapid breathing to dispel CO_2 and thereby reduce acidity, and by boosting the action of the kidneys, which leads to dehydration. If the acidity cannot be corrected by natural means, tissue damage occurs and eventually death.

More than half the people in developed societies die of heart disease. Rather than wait until their heart is showing signs of weakness, when they would be prescribed aspirin by their doctor, many now believe they can escape this fate by the simple expediency of taking a junior aspirin tablet every morning as a preventative. Such a tablet contains a quarter of the normal dose, in other words 75 mg of acetyl salicylic acid. What they may not realize is that they are also getting salicylate from other sources, notably their diet.

Some who fear for the health of their heart have been persuaded that they can fend off the grim reaper by eating the right type of fat. They avoid all animal fats and hydrogenated vegetable oils, and go for those vegetable oils which are mainly mono-unsaturated. They may also have read that those who drink red wine are also less prone to heart disease. All this advice for a healthy life appears to be sound, and those who advocate it can point to the people of the Mediterranean region where heart disease is much less common than elsewhere. Clearly the diet of that region must hold the key, they say, and the focus tends to fall on olive oil and red wine. A chemical explanation is usually offered in terms of mono-unsaturated fats, the main component of olive oil, and polyphenol antioxidants, which are particularly abundant in the skins of black grapes and which are extracted into red wine.

It may well be that the Mediterranean-style diet has another factor: salicylate. This is also present in many vegetables, herbs and fruits. Gazpacho, the soup made from tomatoes, onions and tarragon, and served cold, may contain a healthy dose of salicylate, while ratatouille, the vegetable dish of aubergine, courgettes, red peppers and tomatoes, could be brimming with the stuff. Other foods from warm climes are relatively rich in salicylate, such as pineapples, melons and mangoes, while curry powder has over 200 mg pre 100 g.

It is possible to plan a diet that will garner salicylate in gentle stages throughout the day. For example, if you like fruit at breakfast go for raspberries: a bowl of them will provide 4 mg of salicylate. If you want a salad at lunch then choose chicory leaves and add a couple of gherkins: both have lots of salicylate. A liberal sprinkling of tomato ketchup on your hamburger and fries is also a good idea, and if you need a snack during the day then nibble a handful of currants or raisins.

The easiest way by far to boost your salicylate intake is to drink tea. A cup, made with one tea bag, will provide 3 mg, and if you drink the average 5 cups a day you will be getting a life-enhancing 15 mg. Coffee drinkers, on the other hand, would need to take in 20 mugs of their brew to get this amount. Other foods to boost your daily dose of salicylate are almonds, peanuts, coconut, honey, Worcester sauce, licorice, peppermint, broccoli, cucumbers, olives and sweetcorn. And only eat potatoes with their skins on: peel them and all the salicylate is gone. The same is true of pears. If you are going to a party you can enjoy salicylate in fruit juices, wines and beer.

Of course you may be one of the unlucky few per cent who react badly to salicylate, and are advised to avoid aspirin because it can cause stomach bleeding and ulcers. In which case you are probably likely to get indigestion from a diet rich in salicylate, so the best advice is to avoid foods like these. If you are hypersensitive to salicylate then you may even be put on a salicylate-free diet, but you need not feel deprived because there are lots of zero-salicylate foods to choose from: meat, fish, milk, cheese, eggs, wheat, oats, rice, cabbage, brussels sprouts, celery, leeks, lettuce, peas and bananas have none at all. And if you fancy a drink then stick to spirits, but be careful to choose the right mixer. Gin and tonic is fine, and so is rum and coke, but avoid Bloody Marys (vodka and tomato juice).

Portrait 9

Those unspeakable molecules—phthalates

Finally in this gallery we come to a portrait of a molecule that is present in everything we eat: phthalate. There have been several scares about phthalates over the years: a recent one in the UK concerned their presence in formula feeds for babies. Mothers were alarmed to be told that phthalates were contaminating their baby's feed, and that these molecules were being described, somewhat mischievously, as 'gender-bending' chemicals. The panic that resulted echoed an earlier phthalate scare of the 1970s when they were said to leach from plastic wrapping into food, and were then accused of causing cancer. Despite these worrying assertions, there is no need for alarm, because phthalates cause neither cancer nor infertility in humans, as we will discover.

Phthalates are derivatives of phthalic acid, which consists of a benzene ring with two acid groups attached. These groups may be next to each other, when the molecule is called simply phthalate, or on opposite sides of the ring, when it is called terephthalate. (There is a third form in which the groups are one atom apart, but these have little commercial significance.) Phthalates were first made in the 1850s and called naphthalates, from *naphtha*, the ancient Greek name for natural petroleum, but this was soon shortened to phthalate.

Phthalates are entirely manufactured and worryingly widespread; even in remote regions of the planet analysts have recorded 0.5 ppm of phthalates in rainwater, so even the peoples of the high Himalayas and the remote Pacific islands get a daily dose. The alarm over baby foods came from a report by the UK's Ministry of Agriculture, Fisheries and Food, which released surveys entitled *Phthalates in Paper & Board Packaging* (1995) and *Total Diet Survey* (1996) which reported them to be present in almost all food analysed, not just in baby milk. Levels in milk and milk products were reported to be around 1 ppm, and for a time it looked as though this might be coming from the PVC tubing used in milking machines, but investigation showed that this source accounted for only a tenth of what was present.

Both kinds of phthalate are produced industrially. Terephthalate is used to make polyester for bottles and fibres; it is permanently fixed as an integral part of the polymer and poses no threat. We will be inspecting its portrait in Gallery 5. The other kind of phthalate goes into plastics like PVC to make them pliable. PVC is a tough, rigid solid used for window frames and drainpipes, but when phthalate is added to it the plastic becomes flexible because this allows the polymer chains to move over one another. In this way we get PVC that can be used as garden hoses, wallpapers, shower curtains, clothes, blood bags and water beds. However, it is electric cable and vinyl flooring which uses most of the phthalate. This phthalate is not fixed, and is simply blended in to act as a molecular lubricant. If one of these phthalate molecules finds itself near the surface of the PVC it is free to escape—to be rubbed off or to evaporate into the air.

Because of earlier fears about their safety, plasticizer phthalates are now among the most investigated of all chemicals. The leading plasticizer is DEHP, short for di(ethylhexyl) phthalate, but according to David Cadogan, of the European Union's Council for Plasticizers and Intermediates in Brussels, this poses little risk: 'As far as humans are concerned it causes neither cancer nor reproductive effects. Nor are phthalates accumulating in the environment because they are biodegradable, and levels are falling. In Rhine sediment, for example, there has been a reduction of 85% since the 1970s. Phthalates are very insoluble in water—about a millionth of a gram per litre—so leakage from plastics in old landfill sites is tiny.'

In 1990 the EU Commission said that DEHP should not be classified as a carcinogen, because no carcinogenic or oestrogenic activity was found with fish, hamsters, guinea-pigs, dogs or monkeys. However, rats did show increased risks of liver tumours and smaller testes, but these animals, unlike humans, are known to be particularly prone to respond this way because they have been specially bred to be sensitive to cancer-forming chemicals. Humans are not at risk. The Danish Institute of Toxicology concluded that an intake of 500 mg a day was without effect. Our average daily intake is around 0.35 mg, which over a lifetime would amount to less than 10 g (a dessert spoonful). For babies, the tolerable daily intake is 0.05 mg per kilogram of body weight, but no formula feed would provide anything like this amount of DEHP. In any case the 0.05 guideline has a large inbuilt factor and is based on the tests on rats. The danger from phthalates is negligible, even to babies. If all the phthalates in a year's supply of milk were to be consumed at one feeding, it would still not be enough to make a baby sick, let alone anything more serious.

GALLERY 2

TESTING YOUR METAL

An exhibition of the metals which our body must have

■ BONE IDLE ■ THERE IS NO SUBSTITUTE FOR SALT
■ PERFECT AND POISONOUS ■ THE ENIGMATIC ELEMENT
■ AMAZINGLY LIGHT ■ THE MISSING LINK ■ BRONZED AND BEAUTIFUL
■ A FAMILY PORTRAIT

ASK PEOPLE which *metals* are essential for healthy living and I suspect most would say zinc and iron. Some might mention sodium and potassium, although sodium is often regarded as something deleterious to healthy living; and a few people will know that calcium is a metal also, and important. In fact the human body needs *fourteen* metal elements to function properly.

But for every metal that we do need, there is another that our body contains that we could well do without. These metals serve no known purpose, but they come with the food we eat, the water we drink, and the air we breathe and our body absorbs them, mistaking them for more useful elements. As a result we find that the average adult contains measurable amounts of aluminium, barium, cadmium, caesium, lead, silver and strontium. There are also trace amounts of many others, including gold and uranium.

Because strontium so closely resembles calcium we absorb a lot of this element, and the average person has about 320 mg in their body, far more than of many of the essential elements. On the other hand the weight of gold in the average person is only 7 mg, worth but a few pence, and the weight of uranium is only 0.07 mg, although turned into pure energy this could drive your car for five kilometres. Our body tends to retain these unwanted intruders either in our skeleton, as in the case of uranium which has a special propensity to bind to phosphate, or in our liver which has proteins that can trap metals like gold.

The table below lists the amounts of the essential 14 metals in the average adult—someone who weighs 70 kg (155 pounds). As we would expect, calcium heads the list because, along with phosphate, it is what makes up the bones of our skeleton, which weighs 9 kg on average. Of this, 1 kg is calcium and 2.5 kg is phosphate. In fact 99% of the body's calcium and 85% of its phosphate is in the skeleton. Bone also contains water and the protein collagen, plus the elements sodium, potassium, iron, copper and chlorine. There is also lead, an element that has a particular affinity for phosphate.

Bones can preserve forensic or archaeological data because of the metals they hold. For example, the extent to which earlier civilizations were exposed to lead can be gleaned by analysing their skeletal remains, which sometimes show over 100 parts per million (ppm) of lead. The level in human bones today is about 2 ppm. We will find a portrait of lead in Gallery 8.

At certain stages of our life it is vital to ensure an adequate intake of calcium: when we are growing up, when we are pregnant or breast feeding, and perhaps even as we get old, when our bones lose calcium and weaken. However, there is no strong evidence that a high calcium diet in old age can reduce this loss.

The 14 metal elements essential to the human body

	Metal	*Amount*		*Metal*	*Amount*
1	calcium	1000 g	8	tin	20 mg
2	potassium	140 g	9	vanadium	20 mg
3	sodium	100 g	10	chromium	14 mg
4	magnesium	25 g	11	manganese	12 mg
5	iron	4.2 g	12	molybdenum	5 mg
6	zinc	2.3 g	13	cobalt	3 mg
7	copper	72 mg	14	nickel	1 mg

After calcium the next two most common elements are potassium and sodium, which go hand in hand to operate the electrical signals which transmit nerve impulses to and from the brain. Then come magnesium, iron and zinc, which are needed in amounts that our diet may not always supply. We can also take in too much of them, and in particular of iron. Finally there are the elements which are almost never lacking in our diet, and most of these we will examine as a group portrait at the end of the gallery.

Portrait 1

Bone idle—calcium phosphate

Calcium phosphate is very insoluble and tough, which explains why dinosaur bones have survived for a hundred million years or more. Calcium phosphate is also the reason why soft tissues and leaves can be fossilized as well. Bacteria in conditions where there is no oxygen, such as are found in the mud at the bottom of lakes or swamps, can create microspheres of calcium phosphate within the dead tissues they invade, and it is these which preserve their host's structures in amazing detail. Mineralization of a dead organism takes only a few weeks under anaerobic conditions. Bacteria use the calcium and phosphate of the organism's own cells to construct an outline of microscopic particles of calcium phosphate, thereby preserving it for posterity.

Human skeletons can also be preserved for thousands of years even under normal conditions. In 1994 Mark Roberts of the Institute of Archaeology, at University College London, discovered the oldest bone tools in Europe. These belonged to Boxgrove Man, who lived half a million years ago, near where the East Sussex village of Boxgrove now stands.

We think of bone as inert, but in living things it is constantly degraded and restored, at millions of remodelling sites throughout the skeleton, by cells called osteoclasts and osteoblasts. In this way bone carries out its secondary function of maintaining a steady level of calcium in the blood. This element has several roles to play such as in muscle contractions, cell division, hormone regulation and blood clotting. When our diet does not provide enough calcium for these essential processes, the deficiency is made good from our skeleton store, to be replaced later when there is an excess of calcium in our blood.

As we get older the replacement does not quite compensate for the loss, and so to delay this erosion we need a daily intake of calcium plus vitamin D. This vitamin regulates bone growth, and is abundant in foods such as fish oils and eggs. (As we saw in Gallery 1, we do not need to worry about the phosphate deficiency in our bones because this is never in short supply in our food.) Children who lack vitamin D may suffer from poor skeletal development due to the deficiency disease rickets, and they may end up with bow legs. Calcium supplementation may even benefit normal children, as shown by Conrad Johnston of Indiana University, USA. He took 60 pairs of identical twins, aged 6–14, and gave one of the pair a tablet of calcium every day for three years. These children showed accelerated bone growth.

Those in need of extra calcium will generally get enough from a balanced diet, but it will not hurt to ensure a regular intake of foods particularly rich in this element, such as sardines, eggs, almonds, cheese, milk chocolate and white bread. If you suffer from indigestion and you suck calcium carbonate antacid tablets, you could immediately double your daily intake of calcium, which for the average person should be around 500 mg per day. Those who need more than this are teenage boys, who should aim at 750 mg, teenage girls 650 mg, and breast-feeding mothers 1100 mg. Pregnant women do not generally need extra calcium because their bodies automatically adjust to absorb more of the calcium in their diet, rather than letting it pass through them. On the other hand, pregnant teenage girls do need extra calcium as they have their own growth needs to meet as well as those of the developing foetus.

The weight of calcium phosphate in our skeleton reaches its maximum when we are around 30, thereafter about 1% a year is lost, until in old age our bones become porous and liable to break easily, especially at the hip joint. Doctors who specialize in bone disorders are becoming concerned at the growing number of elderly patients who are admitted to hospitals for hip operations, and it is believed that many of these could be prevented by reducing bone loss in old age and by strengthening peoples' bones earlier in their life.

Women who are in their 50s may suffer massive bone loss during the menopause, although this can be reduced by hormone replacement therapy. More women die from the after-effects of osteoporosis fractures than from all cancers of the ovaries, cervix and womb put together. Certain drugs can slow down calcium leaching, such as calcitonin, a rather costly drug which is given to supplement the natural calcitonin which is secreted by the thyroid gland specifically to counteract bone loss. A synthetic form, salcatonin, works equally well promoting increased bone density after two years' treatment—and those taking it suffer a third fewer bone fractures.

There are other treatments for osteoporosis. For example, hormone replacement therapy will slow down, and may even halt, bone loss. The protective effect of oestrogen on the bone can be mimicked with some compounds known as selective oestrogen receptor modulators, and without unwanted side effects of these drugs. Bisphosphonates also prevent the breakdown of calcium phosphate in the bone, and can actually increase bone mass if taken over a two year period. There is the simple expedient of taking more vitamin D, which regulates bone loss, to counteract the fall-off in the ability of older people to make this molecule by the action of sunlight on the skin.

Fluoride therapy is used in some countries, and given at a rate of 20 mg per day it strengthens bone by infiltrating the calcium phosphate to form a chemical known as fluoroapatite. It has the advantage of being extremely cheap to administer, and seems to work well at preventing bone loss from the spine. Fluoride also strengthens our teeth by the same chemical process. Tooth enamel is a modification of calcium phosphate called hydroxyapatite, which is both stronger and even less soluble than ordinary calcium phosphate. It can be improved even further by the addition of fluoride which converts it to fluoroapatite, a mineral that can then better resist the ravages of the acid attack caused by the bacteria in the mouth which convert sugar to acids.

This discovery of the benefits of fluoride led to the fluoridation of public water supplies, and on 25 January 1945, the citizens of Grand Rapids, Michigan, USA, became the first community to have their tap water artificially fluoridated as a way of improving their children's teeth. Earlier in the century US immigration officials had observed that people from Naples had curiously stained, but very healthy, teeth. Research traced the cause to the Neapolitan water supply, which contained 4 ppm of fluoride. A similar observation was made in the UK in World War II by health authorities, who had to deal with the large-scale evacuation of city children to rural areas where they were safe from air raids. The authorities noted that those children evacuated from the town of South Shields had much better teeth than those from neighbouring North Shields. Again the cause was traced to 1 ppm of fluoride in the water supply of South Shields, which in those days was drawn from artesian wells.

In the years following World War II, many towns and cities in the USA added fluoride to their water supply, and so did a few in the UK. Although in the UK many local authorities have requested fluoridation, water companies are not obliged to comply, and only a few have done so. The water is then treated by adding fluorosilicate, a fluoride-rich by-product of phosphoric acid manufacture.

Fluoridation is seen as the cheapest way of reducing tooth decay, yet despite its benefits, there are those who oppose fluoridation because they want all drinking water to be free of unnatural chemicals. Others are against it on ethical grounds, seeing it as enforced medication, while some even doubt its efficacy, and a few think it is actually harmful. This last group of people cite epidemiological studies that show marginally more cases of bone and liver cancer in fluoridated areas. They also claim fluoride damages the immune system, although this seems most unlikely.

In any case, anyone who wishes can treat themselves by using fluoride toothpaste, which generally has 0.1% fluoride (1000 ppm), with the fluoride being in the form of sodium monofluorophosphate which is a less toxic form of fluoride than sodium fluoride. Such toothpaste is not recommended for young children because they tend to swallow it, and their parents should buy them a toothpaste with 0.05% (500 ppm) fluoride.

Fluoride toothpaste became popular in the 1970s, and it is this which is credited with doing most to reduce dental caries in the UK. Andrew Rugg-Gunn of the Dental School, University of Newcastle-upon-Tyne, England, has researched the effects of fluoridation on children's teeth for almost 20 years, and finds that fluoridation cuts decay by half. In areas where the water supply was not fluorinated there was still a notable decline in dental caries in children once fluoride toothpaste became popular, but eventually there is a levelling off, and further progress can only come about through children having fewer sugary treats and drinks.

Fluoride is chemically akin to chloride, which we consume daily as sodium chloride, or common salt. But while we can sprinkle a few grams of this on our food with impunity, the same amount of sodium fluoride would kill us. At one time sodium fluoride was used as an effective insecticide for cockroaches and ants. Dangerous though fluoride is, we each have about 2 g in our body and we take in about 2 mg a day. A litre of fluoridated tap water has 1 mg, but most people get their fluoride from such foods as chicken, pork, eggs, potatoes, cheese and tea. Cod, mackerel, sardines, salmon and sea salt are particularly rich in fluoride because seawater contains 1 ppm of fluoride. We should not be afraid of fluoride because it is essential for our health: laboratory animals fed on fluoride-free diets failed to grow properly, and were anaemic and infertile. The same would no doubt be true for humans.

If our diet contains too much fluoride we may end up suffering from fluorosis, the first signs of which are mottled teeth, like the people from Naples. Later there may be osteosclerosis, a hardening of the bones, which can lead to a deformed skeleton. In certain parts of India, such as the Punjab, the condition is endemic, especially where villagers drink water from wells with levels of fluoride as high as 15 ppm. About 25 million Indians suffer a mild form of fluorosis, and many thousands show skeletal deformities. In some villages one child in six is affected, but this is improving with de-fluoridation schemes.

Such high levels are rare, and for the rest of us we generally need to increase our exposure to this element, which together with an adequate intake of calcium may ensure us a stronger skeleton and prevent us suffering the dangers inherent in fragile bones when we grow old. The skill is in finding a balance of the two, but this is easily achieved with regular snacks like sardines on toast made with white bread, both of which are good sources of calcium.

Portrait 2

There is no substitute for salt—sodium chloride

Too much salt, or sodium, is bad for us if we have a heart condition, and doctors whose patients are thus afflicted will often recommend a strict low-salt regime. The average person eats about 10 g of salt a day, about three times as much as they really need. A third of this salt comes naturally in our food, but a third comes from salt which is added to prepared foods, such as cereals and bread, while a third comes from the salt we sprinkle on our meals.

Does too much salt really harm us? If we are healthy, then probably not.

Many people like the taste of salt, and many foods, especially snack foods, are liberally dosed with salt. If you are aware of the dangers of salt you will look for packets labelled 'low salt.' You may already be trying to avoid this chemical by using a salt substitute, but you are probably wasting your money. You would be better training your taste buds to enjoy a low-salt way of life, because there is no substitute for salt.

When you read about salt in magazines and books, the words salt, sodium and sodium chloride are often used to mean the same thing. Chemists view these terms rather differently. A salt, any salt, is a compound made up of positive and negative ions, the former generally being metals and the latter non-metals. Sodium chloride is but one salt. Others we may encounter in everyday life are sodium carbonate (washing soda), aluminium sulfate (alum), potassium iodide (used in iodised salt) and calcium phosphate (bone meal). Sodium is a metal element, which in our bodies is present as the positive sodium ion, not necessarily associated with chloride, but free to move independently. It is not wrong to speak of sodium in dietary terms, rather than specifying salt, although salt is the way we ingest most of this essential element.

Sodium chloride triggers a specific reaction on the tongue and is one of the four basic taste sensations. This is rather puzzling because no other salt or mineral in our diet provokes such a response. There can be substitutes for sugar but not for salt, yet many people regard salt in the same way they regard sugar: as pure, white and deadly. There is a grain of truth in this slogan because some medical conditions require a low-salt diet. But even people on such regimes cannot live entirely without a daily supply of sodium.

Every cell of our body needs a little sodium, and some parts of the human system, such as blood and muscles, need a great deal. Sodium is used mainly with potassium to move electrical impulses along nerve fibres. We need sodium and potassium for other purposes too, but this is the most important use of these metals in our bodies. We need to take in a regular supply of them because we lose salt continually from the blood stream as it is filtered by the kidneys and excreted by the sweat glands when they are working to cool us down.

We can recycle some sodium, but even on a salt-free diet we are losing it all the time to the extent of about a gram a day. This is regularly replaced because every mouthful of food we eat contains some sodium, whatever diet we are on, and we never need to add salt to our food to get the one gram a day that our bodies *must* have.

Salt is vital, and our body cares not whence it comes. Some people are prepared to pay a lot for sea salt, using it in place of ordinary salt, but it makes no difference once it reaches our stomach. Others buy no salt at all, but their bodies still get what they need from the food they eat. Still others buy salt substitute, which is a 50:50 mix of sodium chloride and potassium chloride. Both sodium and potassium are essential to life, but we need not plan them as part of our diet because they occur in everything we eat. A glass of beer and a handful of salted peanuts would provide our daily requirement of these two essential minerals; so would a poached egg on toast; and so would a bowl of muesli and milk.

It seems rather odd that sodium chloride gets a bad press while potassium chloride gets a good one. This is the opposite of their behaviour as poisons: you cannot kill yourself by taking a large dose of sodium chloride (it will only make you vomit) or by injecting a solution of it. On the other hand, if you injected a solution of potassium chloride it will kill you within minutes by upsetting the rhythm of your heart. This does not mean that eating potassium chloride in place of sodium chloride, in a low-salt formulation, is risky—it isn't. We need much more potassium chloride in our diet than sodium chloride, but normally we get all we need in the food we eat. We will look a little closer at a portrait of potassium chloride next, but first let us look at sodium chloride.

Salt is called rock salt when it is dug out of the ground, and sea salt when it is obtained by evaporating sea-water until the salt crystallizes. Rock salt is contaminated with sand, and sea salt is contaminated with debris from the ocean. Apart from the dirt, there is no difference. Rock salt is salt from ancient seas, deposited as they dried out millions of years ago; sea salt is from today's seas. Both can be refined by redissolving them in fresh water, filtering off the dirt, and evaporating the filtered solution until the salt recrystallizes as pure sodium chloride. Table salt generally has a little magnesium carbonate added to keep it free-flowing.

A person with an illness like kidney disease may be put on a low-salt, a no-salt, or even a salt-free diet by their doctor. A low-salt diet means you eliminate table salt and the salt used in cooking. Foods high in salt, such as potato snacks and certain cheeses, are also not allowed. Such a diet still provides about 6 g of salt a day from foods like bread and cereals. A no-salt diet cuts out even more foods and limits bread to three slices per day, so that the intake is around 3 g. A salt-free diet means no bread at all, and the aim is to exclude all salt except that which comes as a natural part of the food we eat—the patient has to eat mainly foods which have very little anyway, such as boiled rice.

Excess salt is bad for people with kidney disease and it raises their blood pressure. Could salt be a major factor in regulating the blood pressure of everyone? If it really were true that the more salt we eat, the higher our blood pressure, then clearly salt could be a factor in strokes and heart disease. As we have come to expect with studies based on analysing data collected from questionnaires, the evidence is less than clear cut.

An analysis of several surveys, which together covered 47 000 people, was conducted by a group at St Bartholomew's Hospital, London, in 1992. This study found a link between salt intake and high blood pressure. The conclusion was that it would be possible to prevent 70 000 deaths per year in the UK (population 57 million) by reducing salt in the diet. This headline-grabbing claim implied that salt reduction could save more lives than conventional drug therapy, which many found hard to believe. Could something as common as salt, and so essential an element for the body, really be so dangerous?

At the upper end of the salt scale, there are towns in Japan where the average salt intake is 20 g per day, and yet the average blood pressure among the 40- to 49-year-olds is 143/86. These figures are the systolic/diastolic ratio and they compare the pressure of blood as it leaves the heart (systolic) with the pressure of blood as it enters (diastolic). The lower these figures the better, but they generally rise with age. The upper figure is the one to watch: a rough guide is to keep this at a value that is several points below the total of 100 plus your age, so that for a 50-year-old the danger signal is 150. The Japanese live longer than people anywhere in the world, even though their diet contains a lot of salt (they eat a lot of fish from the sea).

At the lower end of the scale of salt users is a tribe called the Yanomamo, who dwell in the forests of southern Venezuela and consist of an estimated 20 000 people who live by subsistence farming in small villages. They are one of the few remaining tribes unaffected by Western culture. The men spend about three hours a day tending their plots. The rest of the time they devote to planning and carrying out raids on other villages, killing and raping. The Yanomamo eat virtually no salt at all. Researchers observed 46 members of this tribe who were in their 40s, and found they had an average blood pressure of only 103/65. Another Amazonian tribe, the Carajas, take in a little salt, calculated to be half a gram a day, and the average blood pressure of ten of their middle-aged people was slightly lower at 101/69. (The longevity of these people is not recorded, but if there is a link between salt, blood pressure and lifespan then we can assume they will probably all live to be a hundred.)

For the rest of us, it is not easy to control our salt intake because so many food products contain it. Indeed, without salt we would find bread, cheese and breakfast cereals tasting most unusual. Nevertheless, people became so worried by alarms about salt that manufacturers responded, producing foods with reassuring labels that proclaimed them to be 'low salt' and even 'salt free.' Some showed a little more sophistication and said 'low sodium.' Medical research findings, however, are divided on whether salt really is a threat to healthy people.

The Intersalt Cooperative Research Group looked at 10 000 men and women from 52 places around the world and concluded in 1988 that if we cut our average salt intake from the normal 9 g a day to 4 g a day we could lower our systolic blood pressure by 2 units, but our diastolic blood pressure would not change at all. In 1996 the Intersalt group reported more forcefully, and in the *British Medical Journal* their chief investigator, Paul Elliott of Imperial College School of Medicine, London, said that reducing salt in manufactured foods would have a bigger effect on deaths from heart disease than all the drugs used to treat high blood pressure put together.

Meanwhile that year the *Journal of the American Medical Association* published a report from the Mount Sinai Hospital, Ontario, which came to a different conclusion. This group, led by Alexander Logan, found that salt posed no threat at all to people whose blood pressure was normal. The only case for restricting sodium intake was with older people who already had high blood pressure.

When experts differ like this, lay people tend to be rather sceptical, wondering if the dietary salt debate will ever be resolved one way or the other. The only advice worth acting upon is that of one's doctor, and if he or she tells you to cut down on salt, you would be foolish not to. If you have the major factors which affect health well under control, in other words you do not smoke, are not overweight and you drive carefully, then it might be worth considering dietary advice about salt. Even so, the major factor in longevity is to have the right parents, but this is not something we have much control over.

Although salt and sugar have been much abused as typical ingredients of junk foods, they can be a blessing in tropical countries, where they save millions of lives. Diarrhoea, and the dehydration it causes, kills twelve million children each year. The answer to this fatal illness is not expensive antibiotics, but sugar and salt. A drink made of eight teaspoons of sugar and one of salt in a pint of water will save a sick child's life. Sugar and salt may be junk food components, but working together these two chemicals can restore the lost bodily fluid, cheaply and effectively.

Portrait 3

Perfect and poisonous—potassium chloride

While many people regard sodium as something to be assiduously avoided, they will probably hold a very different view about the companion metal potassium, and may even actively seek out foods that are rich in this element in the belief that the more they take of it the healthier they will be. This is not a bad idea, because we need a lot more potassium than sodium in our diet, and we do have 40% more potassium in our bodies than sodium. The average adult contains about 140 g (5 ounces) of potassium, but only 100 g (3.5 ounces) of sodium. This ratio is reflected in the recommended daily allowances for these two elements, which, for adults, are 3.5 g a day for potassium and 1.5 g for sodium.

Almost all food contains potassium except vegetable oils, butter and margarine. Potassium is essential, and some foods are particularly rich in it, such as seeds and nuts which may have up to 1% by weight of this metal. A more normal range is 0.1–0.4%. Common foods with over 0.5% potassium are kippers, peanuts, raisins, potatoes, bacon and mushrooms. Some foods have more than 1% potassium, such as All-Bran (1.1%), butter beans (1.7%), dried apricots (1.9%), yeast extract (2.6%) and instant coffee (4.0%).

Potassium is found in all parts of the body. Red blood cells have most, followed by muscles and brain tissue. It is found mainly in the fluid between the cells of the body as an electrolyte, and its most important role is in operating our nervous system. Positively charged potassium and sodium atoms move in and out of channels in the membrane of nerve cell walls, but some channels only permit potassium to pass through. The result of all this activity is equivalent to an electric current passing along the nerve fibre.

The black mamba snake has developed a venom that works by blocking these channels, thereby preventing impulses from flowing along the nerves, so that the victim goes into convulsions and dies. The toxin of the black mamba has been used to probe the layout of the potassium channels in the human brain by injecting volunteers with tiny amounts of a radioactive form of the venom, and then mapping where it has collected. This research shows that the highest concentration of such channels occurs in the hippocampus region, the part of the brain that is important in learning. (As yet no one seems to have put forward a theory that a high potassium diet might help students to pass examinations.)

There are some conditions that lead to potassium deficiency, such as starvation, kidney malfunction, and the use of certain diuretics—drugs given to increase the excretion of urine. We need a constant supply of this essential element to make lean tissue and to keep our kidneys working. If we are not getting enough potassium we experience muscular weakness, and this has an effect on the heart muscle, causing irregular beats and even cardiac arrest. Chronic potassium deficiency leads to depression and confusion.

Medical treatment sometimes requires potassium supplements, and some medicines like diuretics may have extra potassium included in their formulation. However, it is rare for potassium deficiency to lead to ill health because it is almost impossible not to absorb potassium from the food we eat, particularly from vegetables and fruits. All plants absorb a lot of potassium from the soil, and it was the ash of wood fires, so-called potash, which gave this element its name. Heavy beer drinkers may get too much potassium, and their craving for salted snacks may be the body's way of maintaining the sodium–potassium electrolyte balance.

A chronic excess of potassium in the body depresses the central nervous system. Large doses of several grams of potassium chloride will paralyse the central nervous system, and cause convulsions, diarrhoea, kidney failure, and even heart attack. Another way of disrupting the body's potassium balance is to inject a solution of potassium chloride, and this can be fatal. When there is too much potassium outside the nerve cells, the potassium inside the cells finds it impossible to escape, and the electrical impulse they should be transmitting dies away. All body functions are affected, but none more dramatically than the heart, whose muscles stop beating.

A British doctor, Nigel Cox MD, devised this method of killing one of his terminally ill patients, 70-year-old Mrs Lillian Boyes. He dispatched her with an injection of potassium chloride solution, but was later arrested, tried, and found guilty of murder. It came as a surprise to many that a simple salt, which is essential to all living things and is openly sold on supermarket shelves as a salt substitute, could be lethal. Strangely, what Dr Cox did illegally in the UK is done legally in the USA as a form of capital punishment in certain states. Condemned men who agree to donate their organs for transplants may be executed by being given what is described as a non-toxic lethal injection of potassium chloride. This chemical kills, but, unlike poison gas or the electric chair, it leaves all the organs undamaged.

The world production of potassium ore is about 40 million tons, mainly from mines in the UK, Germany, Canada and Chile, and from Dead Sea brines. The Chilean deposit is potassium nitrate (saltpetre), but the other sources are potassium chloride mixed with salts such as sodium chloride and magnesium chloride. In the UK there is a potassium chloride mine over a kilometre deep which produces almost a million tons a year of the pinkish ore sylvinite. The process of extraction involves crushing the ore and then separating the potassium chloride from the other minerals present by using a concentrated solution of sodium chloride of exactly the right composition to ensure that no potassium chloride dissolves. The bulk of the potassium chloride is used for making fertilizers, while the rest goes into chemicals, such as potassium hydroxide, used in liquid soaps and detergents, and potassium carbonate, used in special glass for television sets. A little potassium chloride goes into pharmaceuticals, hospital drops and saline injections.

A rather unusual use for potassium chloride has been suggested, and that is to increase rainfall over regions prone to drought. Normally clouds release only about a third of their moisture as rain, but this can be doubled if they are seeded with fine particles. Even a 10% increase in rainfall in some regions could be of tremendous benefit to farmers. Graeme Mather of South Africa invented the new method of seeding clouds that uses flares mounted on the wings of aircraft. These fly beneath the clouds and release a smoke of potassium chloride, which then drifts up into the cloud—and down comes heavy rain. Independent tests by the National Center For Atmospheric Research in Boulder, Colorado, in 1995 showed that the method worked.

Portrait 4

The enigmatic element—iron

It is common knowledge that without iron we become anaemic, but keeping our red blood cells working efficiently is only one role that iron plays in the body, albeit the most important one. Without iron, blood corpuscles cannot extract oxygen from the air in the lungs, and so cannot distribute it around the body to generate the warmth that keeps us alive. The other roles for iron are in enzymes: those involved in the synthesis of DNA; those which enable cells to release energy by using glucose; and those which scavenge free radicals and protect us. (Iron may also be responsible for actually forming some free radicals.) Normal brain function needs iron, and there are regions of the brain that are rich in it—which may explain why iron deficiency in infants and children has been associated with slower mental development.

Iron is essential to almost all living things, from micro-organisms to humans. We must have a regular intake of iron because we lose a little of this metal every day through the walls of our stomach and intestines. Even so, it is rare for normal people to be lacking iron, even though they may sometimes lose a large amount of blood. If anything, we tend to take in too much iron by supplementing our diet with it, in the popular but mistaken belief that it will keep us feeling alert and full of energy.

The average man needs an intake of 10 mg of iron a day, and the average woman 18 mg, but the amount in food is generally sufficient to provide all that is needed. Women require more of this metal when they are pregnant, and then they should eat iron-rich foods such as liver, corned beef, iron-fortified breakfast cereals, baked beans, peanut butter, raisins, bread, eggs, curries and cakes made with black treacle (molasses). If none of these rather plebeian foods looks appetising, you could eat a more patrician diet of caviar, venison and red wine, all of which have a great deal of iron.

The uptake and loss of iron from the body is finely balanced. We want to take in enough iron to replace that which is lost, but we do not want too much as it can be dangerous. The average diet provides about 20 mg of iron, which seems more than ample, but we need to take in a lot because only about 2 mg of the iron we ingest can be absorbed into our bloodstream. Most of the iron in the food we eat passes straight through us because it is either the wrong form of iron, or it is bound up with the indigestible parts of our food. The amount which is absorbed balances the amount we lose each day, which is about 2 mg. It does not require much of a change in these figures for the body to suffer iron deficiency or iron overload. The former is the more likely, and this is reflected in the estimated 500 million people in the world who are thought to be anaemic. This problem is likely to get worse because the high yield crops that have been developed to feed the world do not provide much in the way of dietary iron.

Inside the body iron is strongly bound by transferrin, a protein found in serum and other secretions, and it is this which transfers the metal between cells. Transferrin binds iron tightly, and because it does so it acts as a powerful antibiotic simply by preventing iron falling into the wrong hands, in other words being available to invading bacteria. These need iron if they are to multiply. Mother's milk also contains a form of transferrin called lactotransferrin, and egg white contains ovotransferrin, both of which function in the same iron-binding way, and both of which provide anti-bacterial protection as a result.

Some people retain too much iron and suffer iron overload, which can have a devastating effect. Increased levels of iron in the brain have been identified as part of degenerative conditions such as Parkinson's disease. Individuals who have the genetic disorder haemochromatosis absorb too much iron, which concentrates in the pancreas, liver, spleen and heart and interferes with their normal functions. Patients who need multiple blood transfusions for inherited diseases such as anaemia, and especially those with thalassaemia, also accumulate too much iron. In the past those with this condition suffered an early death because of iron overload, although now they are given chelating drugs that can help the body shed its burden of iron.

Excess iron in the body may also lead to increased risk of cancer. Surveys have shown that those with transfusion-dependent thalassaemia, and those who work in iron mines and iron foundries, have higher incidences of cancer than normal. In Russia, cancer has for many years been referred to as a 'rusting' disease, perhaps recognizing that iron may have a role to play in this condition. However, it is still not clear what this role could be, other than iron's ability to generate free radicals that can trigger the cancer.

Although iron is the fourth most abundant element in the Earth's crust (coming after oxygen, silicon and aluminium), there are regions of the planet where it is so lacking that it is the limiting factor to life. This is especially true of the surface layers of large stretches of the oceans, which are more devoid of life than any desert on land. We may think of the seas as teeming with life, but this is true of only a few regions, and these tend to be over-fished as a consequence. More than 80% of the boundless ocean is empty. In the mid-1980s, John Martin of the Moss Landing

Marine Laboratories, in California, put forward the theory that it was the lack of iron in the upper levels of the sea which prevented plankton from growing, and without plankton to feed on, other forms of marine life have no food supply to support them.

In the mid-1990s Martin's idea was tested by a joint US–UK research team, which used a solution of iron sulfate to fertilize 60 square kilometres of the Pacific Ocean, West of the Galápagos Islands. The results were dramatic. Within a week this barren span of ocean bloomed and turned green with plankton, proving that it was simply a lack of iron that was limiting their growth. Suddenly it was realized that we could fertilize the oceans just as we fertilize the land, and we could do this with ferrous sulfate, which is easily made from any rusty old iron. The seas would bloom, becoming a marine wonderland as zooplankton and higher forms of marine life fill the empty oceans. By taking most of our protein as fish from the sea, instead of as meat from the land, we could return vast areas of farmland to the wild.

Portrait 5

Amazingly light—magnesium

People often express surprise when they are told that there is almost ten times as much magnesium as iron in the human body. What is this curious metal doing? We may have seen our chemistry teacher at school set fire to a strip of magnesium ribbon to demonstrate how easily the metal will burn and how bright is the light that it produces. This was why magnesium fire bombs in their millions were once scattered over British, German and Japanese cities to set them ablaze, and why magnesium flash bulbs were used in photography. Today society has better ways of using this remarkable metal, which also has a crucial role to play in our bodies.

We begin to consume magnesium with our mother's milk, and we need a daily supply of this metal for a healthy body. Could it be that some of us even lack magnesium in our diet? In 1991, in the medical journal *The Lancet*, Dr Mike Campbell suggested using magnesium salts as a treatment for chronic fatigue syndrome, the puzzling condition referred to as ME (myalgic encelphalomyelitis) by those who suffer from it, and called yuppie 'flu by those who wish to mock it. Campbell reported the results of a double-blind test on 30 ME patients—that is, a test in which neither the patient nor the nurse knows whether the test material or a placebo is being used. Half the patients were given weekly injections of a solution of magnesium sulphate for six weeks, while the other half were given injections of distilled water. Twelve of those on magnesium responded positively, compared to only three of those given water. Patients on magnesium therapy reported having more energy, feeling better and coping more easily with pain.

Whether or not a lack of magnesium really is a factor in ME remains to be seen, but in any case magnesium deficiency is extremely rare. We need about 200 mg a day of this metal, but our body deals very efficiently with magnesium, taking it from our food when it can, and recycling what we already have when it can't. Our daily intake ranges from 350 to 500 mg (about a fiftieth of an ounce). Excess magnesium is not easily absorbed by the body, and too much acts as a mild laxative, as we discover when we take Epsom salts (magnesium sulphate) or Milk of Magnesia (magnesium hydroxide).

Magnesium is certainly an essential element for all living things because it is at the heart of the chlorophyll molecule. Plants need this to trap the energy of sunlight in order to make sugar and starch. This planet is green because magnesium-chlorophyll abstracts the blue and red of sunlight and reflects the green. Plants take their magnesium from the soil, and we take magnesium directly by eating plants, or indirectly by eating the animals that feed on them.

Magnesium disperses throughout the body, with most going into our bones, which act as a magnesium store. Magnesium has three functions: it regulates movement through membranes; it is part of the enzymes which release energy from food; and it is used for building proteins. We rarely need to worry about getting enough magnesium, but cases of magnesium deficiency do sometimes occur, generally as a result of malnutrition, alcoholism, or old age. Lack of magnesium manifests itself as lethargy, irritation, depression and possibly personality changes, all of which symptoms are thought to typify ME.

A normal diet provides more than enough magnesium and most foods contain it, but spirits, soft drinks, sugar and fats contain virtually none. As we saw in the previous Gallery, rhubarb (pieplant) and spinach actively prevent magnesium from being absorbed because the oxalic acid they contain forms a compound that we cannot digest. But do not worry if you enjoy eating these foods, because they are hardly likely to be the only source of magnesium in your diet. Cooking does not affect magnesium, although if you throw away the water in which green vegetables are boiled, you discard over half of their magnesium. Some foods have high levels of magnesium, such as almonds, Brazil nuts, cashews, soybeans, parsnips, bran, chocolate, cocoa and brewer's yeast. All have more than 200 mg per 100 g (i.e. 0.2%).

Dr Eric Trimmer in his book *Magic of Magnesium* argues that modern diets interfere with the absorption of magnesium by increasing our intake of phosphate. The combination, magnesium phosphate, is highly insoluble. His book also claims, rather surprisingly, that magnesium can alleviate premenstrual syndrome and combat osteoporosis. It recommends boosting the magnesium content of our diet by focusing on bran, cocoa and Brazil nuts, all of which contain more than 0.4% of the metal.

Magnesium is the fifth most abundant metal on the surface of the Earth (after aluminium, iron, calcium and sodium), and there are mountains of ores such as dolomite (magnesium calcium carbonate) and carnallite (magnesium calcium chloride). Magnesium salts are slowly leached from the land by rivers and carried to the sea, which explains why seawater is 0.12% magnesium, with the oceans holding a *million* trillion tons of it. Commercial production of the metal now exceeds 300 000 tons a year, with about half this amount extracted from seawater. Most of this magnesium is used to remove sulfur in steel-refining and to alloy with aluminium to strengthen it. There is an expanding market for equipment made from magnesium metal itself, and although magnesium is renowned for its fiery power, it does not burn as the bulk metal and magnesium tubes and rods can safely be welded.

In 1990, in the Tour de France cycle race, the leader of the Dutch cycle team, Phil Anderson, rode a bicycle made with a pure magnesium frame. This gave a better combination of strength and lightness than a steel frame, which is nearly five times heavier, and even than an aluminium frame, which is one-and-a-half times as heavy. A complete magnesium frame can be cast as a single component, so avoiding welded joints while maximizing lightness and strength.

Magnesium is used for luggage frames, disc drives and camera parts, where again lightness is important. Production of this versatile metal is expected to exceed 500 000 tons per year early in the next century, as carmakers discover the environmental benefits of magnesium for lighter and longer-lasting vehicles. Mercedes already uses it for seat frames and Porsche for wheels. Lighter cars means lower fuel consumption and less danger if they are involved in a collision.

Portrait 6

The missing link—zinc

Some food advisors think zinc may be in short supply in the modern Western diet, and there is a growing realization that we are more likely to be deficient in zinc than in iron. It may well be that, in future, breakfast cereals could be fortified with zinc as well. Zinc plays a part in running the body in two ways: it is present in many enzymes, and more than a hundred kinds have been identified; and it is part of the proteins that act as transcription factors, and these appear to be even more numerous. (Transcription is the process whereby RNA is synthesized from a DNA template.)

Zinc's presence in enzymes was the first of its roles to be recognized, and it is at the heart of many of them—especially the enzymes which regulate growth, development, longevity and fertility. People living in parts of the world, especially in the Middle East and Egypt, where the amount of zinc in the soil is low, may exhibit deficiency symptoms such as stunted growth through lack of this metal.

The importance of zinc for certain enzymes was discovered earlier this century, but it was not until 1968 that the first cases of human zinc deficiency came to light, in Iran. There Dr Ananda Prasad was puzzled by one of his patients, a 21-year-old man, who had the weight and sexual development of a boy of ten. His diet consisted mainly of unleavened bread, milk and potatoes. Nor was he the only example, and Prasad found others with the same stunted growth. From his knowledge of zinc deficiency and its effect on animals, the doctor guessed that these men were suffering from the same condition, although previously this had never been diagnosed in humans. When Prasad moved to Egypt he began to research the problem more deeply, taking as his subjects the young men who had been rejected for army service because of their dwarf-like stature. Using double-blind tests with zinc sulfate or placebo supplements, he was able to demonstrate that zinc deficiency really was the cause, and in 1972 he published his results. Prasad became the acknowledged world authority on zinc metabolism, and he wrote a book, *The Biochemistry of Zinc*, which became the definitive work on the subject. The book also includes an account of the genetic disorder acrodermatitis enterophathica, which was previously considered to be fatal for all babies born with it, but which can now be cured with doses of zinc.

For most people, zinc intake is not a problem. The average adult contains around 2.5 g, mainly in muscle tissue, and takes in anything from 5 to 40 mg per day, depending generally on their liking for meat. The intake for men should be around 7.5 mg per day and for women 5.5 mg. Beef, lamb and liver have some of the highest zinc levels, as do oysters, herrings, and most cheeses. Strict vegetarians can get their zinc from sunflower or pumpkin seeds, brewer's yeast, maple syrup and bran.

Levels of zinc are highest in the prostate, muscles, kidney and liver; semen is particularly rich in zinc. Some evidence suggests that for a few individuals, even in the West their diet may not supply enough of this metal, and the lack of zinc could be responsible for the low sperm counts in men observed in certain regions. Oysters are one of the foods richest in zinc, and a pound of them provides 120 mg. The great lover of the eighteenth century, Casanova, regularly feasted on them, although he was not to know that they were replenishing his zinc levels, which might well have been depleted by his infamous activities.

Zinc also plays a role in the way our body deals with alcohol. This is broken down in the liver by an enzyme called alcohol dehydrogenase, which has a zinc atom at its centre. Excessive consumption of alcohol damages the liver, and it has long been known that those with cirrhosis have lower than normal levels of zinc in that vital organ. Again, if the damage is not severe, then zinc supplements can restore the liver's function.

Happily, zinc salts are not toxic to humans, and they can be purchased over-the-counter in many health shops and pharmacies. They are often recommended by dietary advisers for conditions on which standard treatments seem not to work, such as anorexia nervosa, premenstrual tension, post-natal depression, acne and the common cold. If these conditions are exacerbated by the lack of zinc, then clearly they could respond to the taking of zinc supplements.

What should the average person do about their intake of iron, magnesium and zinc? The one to worry about if you are a woman is iron, and zinc if you are a man. Both sexes might find it useful to heed their grandmother's advice and start the day with a bowl of a bran-based cereal, such as All-Bran. While foods like this are still recommended for their fibre content, they also provide a lot of useful metals, as the table below shows.

Bran is the protective coat, known as the husk, which surrounds cereal seeds. It is generally removed when making flour because it consists mainly of cellulose which our body cannot digest, although it acts as the fibre that keeps us regular. The table gives the recommended daily

The essential metals in bran and All-Bran

Metal	Bran[1]	RDA[2]	All-Bran[3]
potassium	1160	3500	480
magnesium	520	300	88
calcium	110	700	45
sodium	28	1600	36
zinc	16	7.5, 5.5[4]	2.5
iron	13	9, 15[4]	3.5
copper	1	2	0.4

[1] mg per 100 g. [2] Recommended daily amount for the average adult. [3] Recommended 40 g serving. [4] The first figure is for men, and the second one is for women.

amounts for the seven most important metals required by the body, and also the amount you would get if you had a bowl of All-Bran for breakfast. This would usually be eaten with semi-skimmed milk, which would boost the amount of some of these minerals such as sodium and calcium. Although it is named All-Bran, this breakfast cereal is really three-quarters bran. Bran by itself would not make a pleasant breakfast cereal, whereas cooked with sugar, and flavoured with malt and salt, it produces a tasty and nutritious start to the day.

Portrait 7

Bronzed and beautiful—copper

If you look back to the table on page 30 you will see that there are eight essential metals of which we have only a few milligrams in our body. These are copper, tin, vanadium, chromium, manganese, molybdenum, cobalt and nickel. There is no danger of our not getting enough of each of these in the food we eat; indeed the problem is that we take in too much, and they have an antagonistic effect on other metals in the body. Copper is one such metal, and high intakes cause serious problems.

Copper is needed for enzymes that are important to our ability to use oxygen effectively. There is no danger that our diet does not provide enough copper, which is not only abundant in certain foods, but also may come with our drinking water if we live in a soft water area and are supplied through copper pipes. In fact, it has been suggested that we get too much copper, and that this works against the iron and zinc in our bodies because copper can displace these metals from their active sites.

In 1818 Christian Friedrich Bucholz detected copper in vegetable ash. In 1850 it was detected in seaweed collected near Saint-Malo, France. But perhaps the most surprising discovery was that in 1847 by E. Harless, who discovered that octopus and snail blood, which is blue, contained copper. We now know that spiders also have blue blood, and all these creatures use the copper atoms in haemocyanin to carry oxygen around their bodies whereas mammals rely on the iron atoms of haemoglobin to do the job.

Copper is essential to all species, but it can be toxic and as little as 30 g of copper sulfate (1 ounce) has been known to kill. This once common component of children's chemistry sets has now been outlawed. However, we are not likely to be poisoned by copper sulfate, because a large dose will act as an emetic and we would quickly vomit it back.

We need to ingest around 1.2 mg of copper per day, and breast-feeding women need around 1.5 mg. We take in copper best from meats as a copper-protein. The foods richest in copper are oysters, crabs and lobster, lamb, duck, pork and beef (especially liver and kidney), almonds, walnuts, Brazil nuts, sunflower seeds, soybean, wheat germ, yeast, corn oil, margarine, mushrooms and, of course, bran. The amount of copper in our diet varies considerably between 0.5 and 6 mg a day, depending mainly on how much of the above foods we eat. Copper concentrates mostly in our liver and bones, and the average person has about 72 mg in their body.

Crystals of native copper occur naturally, and these were presumably the source of the copper beads made by the people of northern Iraq around ten thousand years ago. Pure copper objects are associated with the earliest dynasties of ancient Egypt. Smelting of copper ores began about seven thousand years ago, in the same region, but copper only became a key metal in human development when it was exploited on a large scale as the alloy bronze. The Bronze Age started around 3000 BC and lasted to around 1000 BC, although these times varied considerably from one ancient civilization to another. How bronze was discovered we do not know, but the peoples of Egypt, Mesopotamia and the Indus valley were certainly familiar with it around 3000 BC.

Copper is not difficult to win from its ores, but it is relatively soft. Only when it is hardened by the addition of tin, in the ratio of two parts copper to one part tin, does it form the harder alloy known as bronze. The word copper is derived from the Roman name *Cuprum* for Cyprus, which had been the main exporter of the metal long before it became part of the Roman Empire.

The main copper ore is a yellow copper-iron sulfide called chalcopyrite, which today is mined in the USA, Zaire, Canada, Chile and Russia, and accounts for around 80% of the world's copper (with silver and gold as by-products). A more famous copper ore is the green malachite, which is used for polished slabs, tables and columns, and this is mined in various countries. World production of copper amounts to six million tons a year, and exploitable reserves are expected to last for only another 50 years. This may be unduly pessimistic because less and less copper is likely to be needed for communications networks as optical fibres of glass replace copper wires.

Copper is ideal for electrical wiring because it is easily worked and can be drawn into fine wire, and it has a high electrical conductivity. It also conducts heat very well, and was once used to make pans, kettles and boilers. Traditionally copper was known as one of the coinage metals, and along with silver and gold it was the basis of the currencies of the old world. Because it was the most common metal it was naturally the least valued of the group.

Copper is resistant to air and water, and is used as roofing for public buildings, where it slowly weathers to an attractive green surface patina of copper carbonate.

Portrait 8

A family portrait—tin, vanadium, chromium, manganese, molybdenum, cobalt and nickel

These metals may be needed only in quantities so tiny that a lifetime's supply would be less than 30 g (an ounce). They are deemed essential because they have been shown to be an integral part of various enzymes isolated from animals, and which catalyse processes that are also known to be part of the human system. It seems reasonable to assume them to be just as essential to us. Since the amounts we need are tiny, it generally means there is little risk of us not getting enough of them. We shall look briefly at each one, and in the order in which they are most abundant in the body as listed in the table on page 30.

Tin

The average person weighing 70 kg has around 20 mg of tin in their body, some of which comes from the metal containers of canned foods. Our daily intake is around 0.3 mg, but there is no evidence of humans lacking, or indeed needing, this metal. It may be essential to some creatures, and rats fed on a tin-free diet failed to grow properly, suggesting it has some essential role. The same may be true for humans.

Tin was known to ancient civilizations. A tin ring and pilgrim bottle have been found in an Egyptian tomb of the eighteenth dynasty (1580–1350BC). Tinplate, iron covered with a protective layer of tin, was first mentioned by Theophratus in 320BC. While tin is an Anglo-Saxon word, the chemical symbol for tin, Sn, arises from its Latin name *stannum*. This is also where the word stannery, meaning tin mine, comes from. Tin was traded around the Mediterranean by the Phoenicians who obtained it from Spain, Brittany, the Scilly Isles and Cornwall. Julius Caesar mentions British tin in his book *Commentaries on the Gallic War*.

There was a large tin-plate industry in the Middle Ages in Bohemia and Saxony. Early in the nineteenth century it was discovered that food, and particularly meat, that was sealed in a can made of tin plate would last a long time. Eating such meat, however, could be deadly, as we shall discover in Gallery 8.

The toxicity of most forms of tin is low, but there is evidence that some organotin compounds can have carcinogenic and mutagenic effects in animals. In these compounds tin is chemically bonded to carbon, and they are toxic. Some organotin compounds are used as anti-fouling paint for ships and boats to prevent barnacles growing on them, but even at low levels these compounds are deadly to other marine life such as oysters, and are being phased out.

Tin is a soft pliable metal, but it is not used as such because below 13 Celsius it slowly changes to a powder—an effect known as the plague of tin. However, it appears to be a myth that Napoleon's Grand Army froze to death on its retreat from Moscow in 1812 because all the buttons on the soldiers' uniforms were made from tin which disintegrated in the cold. Today tin is used to plate steel to make cans, and for soldering. The chief ore is cassiterite (tin oxide), which is mined in Malaysia, Sumatra, Russia, China, Bolivia and Zaire. World production is 165 000 tons per year, and although known reserves will last only about 30 years at the current rate of consumption, it is possible to recycle tin from its main use, tin cans.

Vanadium

Like tin, there is a lot more vanadium in the human body than it appears to need. The average person weighing 70 kg contains around 20 mg, and has an intake of around 2 mg a day. Vanadium is thought to be a regulator of one of the enzymes which govern the way sodium operates in the body, but it may have other roles as well. It first aroused the interest of nutritionists in 1977, when some commercial preparations of adenosine triphosphate (ATP) were found to upset the sodium–potassium balance that works the nervous system, and which the ATP was supposed to be acting upon. ATP is the high-energy molecule present in every cell that drives many metabolic processes. The cause of the imbalance was traces of vanadium. So began an upsurge in interest in this element, and while it is still not clear why it is essential in humans it is nevertheless thought to be so. Feeding tests on chickens and rats show that vanadium has a growth-promoting effect, and the same is likely to be true for humans. However, we are not likely to be short of vanadium.

Vanadium is a shiny, silvery metal used mainly as alloys, especially in steels. It was first discovered in 1801 by Andrés Manuel del Rio at Mexico City, Mexico, and then rediscovered in 1831 by Nils Gabriel Selfström at Falun, Sweden. Although there are lots of ores rich in vanadium, it is not mined as such, but generally obtained as by-product of other sources, and from Venezuelan oils. World production is 7000 tons per year.

Chromium

The average person weighing 70 kg has around 14 mg of chromium in their body. It has been shown that animals which lack chromium have an impaired ability to use glucose, suffer mild diabetes, and have reduced cholesterol levels. The same might well be true of humans, but chromium has not been proved to be essential, although there are a few cases on record of people suffering chromium deficiency. It has been observed with Americans that there is a steady fall in chromium in the body with age, but what this means is not known. If it were to be shown to be deleterious, then diets could be supplemented with high-chromium foods such as brewer's yeast, molasses, wheat germ, and kidney. (It would not be wise to supplement our diet with inorganic chromium salts as these can be highly toxic.)

The foods which contain most chromium are oysters, calf's liver, egg yolk, peanuts, grape juice, wheat germ, and black pepper. The average daily dietary intake ranges from 10 mcg to 1200 mcg (1.2 mg), while daily excretion has been measured in the range 50 to 200 mcg. Chromium is poisonous by ingestion, and as little as 200 mg is dangerous. It is also a suspected carcinogen, and the compounds known as chromates have a corrosive action on skin and tissue.

Chromium is mined mainly as the black iron-chromium mineral chromite in Turkey, South Africa, Zimbabwe, Russia and the Philippines, and world production is around 20 000 tons per year. Chromium is a hard, blue-white metal that can be polished to a high shine and resists oxidation and corrosion in air. Its main uses are in alloys, chrome plating and metal ceramics. The name is derived from the Greek *chroma* meaning colour, because its salts are often highly coloured. The pigment chrome yellow (lead chromate) was once popular with painters because of its striking brilliance.

Manganese

The average person weighing 70 kg has around 12 mg of manganese in their body. All plants and animals need this metal but its exact function is still unclear, although it has been demonstrated to be involved in glucose metabolism and in the operation of vitamin B_1, and it is associated with RNA. In 1931 A. R. Kemmerer and co-workers proved manganese to be an essential requirement of mice and rats, and in 1936 it was shown that the disease perosis in chickens could be prevented by given them manganese. Manganese compounds are added to fertilizers and animal feedstuffs, because it may be lacking in certain soils and animals grazing on such land may suffer manganese deficiency.

The daily dietary intake averages 4 mg, but may range from 1 to 10 mg, the upper value of which is not far off the dangerous intake of 20 mg. However, the toxicity of manganese very much depends on its chemical form. The divalent manganese ion (Mn^{2+}), which is how we normally take it in, is not very poisonous, but the permanganate ion (MnO_4^-) is very toxic. Manganese compounds are experimental carcinogens and teratogens. Few poisonings have been caused by ingesting manganese compounds, but exposure to dust or fumes is a health hazard and working conditions should not exceed 5 mg/m^3 even for short periods. Workers breathing in the fumes from manganese metal can be affected with so-called 'fume fever', which manifests itself as fatigue, anorexia and impotence,

There is never any need for humans to take manganese supplements because we get more than enough of this element from our diet. Foods rich in manganese are sunflower seeds, coconuts, peanuts, almonds, Brazil nuts, blueberries, olives, avocados, corn, wheat, bran, rice, oats and tea. The French delicacy snails also contains a high level.

World production of manganese is in excess of six million tons per year, coming mainly from South Africa, Russia, Gabon and Australia.

There are estimated to be a trillion (10^{12}) tons of manganese nodules on the floor of the Pacific Ocean, but if these were to be harvested then it would be for the copper, nickel and cobalt which they contain, rather than the manganese. Manganese itself is not used as such: 95% of mined ore is used to produce alloys, mainly with nonferrous metals such as copper and aluminium; and the other 5% is used to produce manganese compounds. The addition of manganese improves the strength, working properties and wear resistance of steel.

The colour of true amethysts is due to traces of manganese. The mineral pyrolusite, which is manganese dioxide, was used by glass makers in the Middle Ages to remove the greenish tint of natural glass—a small amount added to the molten glass gave a crystal clear material. If more was added then it would colour the glass purple.

Manganese was one of the first metals to be found in all living things. Even in the eighteenth century it was demonstrated that when hydrochloric acid was added to the ashes from a log fire, chlorine gas was given off—a sure sign that manganese dioxide was present, because this chemical can release chlorine gas from hydrochloric acid. It was also observed that the same happened with the ashes of other plants. In 1808 manganese was found in ox bones, in 1811 in human bones, and in 1830 in human blood.

Molybdenum

The average person weighing 70 kg has around 5 mg of molybdenum in their body, but this amount taken in a single dose would be dangerous, and 50 mg is sufficient to kill a rat. The average human takes in about 0.3 mg a day (which is around 8 g, or about a quarter of an ounce, in a lifetime). Foods which have the most molybdenum are pork, lamb and beef liver, sunflower seeds, soybeans, lentils, peas and oats.

Despite its toxicity molybdenum is essential to all species. There is a mammalian enzyme, xanthine oxidase, in which molybdenum is present and which is a vital constituent in the production of uric acid, which is how we excrete unwanted nitrogenous material from the body. If this enzyme is too active it can lead to gout, which is the painful accumulation of sharp crystals of uric acid in the joints. Modern treatment targets the enzyme to depress its activity.

There are 20 molybdenum-containing enzymes, the best known of which is the nitrogen-fixing enzyme nitrogenase found in nodules in the roots of certain plants such as legumes, as we shall see when we come to inspect the portrait of nitrogen in Gallery 6. Another molybdenum enzyme is needed in the metabolism of alcohol. This chemical is first converted by a zinc-containing enzyme to acetaldehyde, which is converted by a molybdenum-containing enzyme to acetic acid, which is then used as a source of energy and converted to carbon dioxide. Some races, such as the Japanese, have low levels of the molybdenum enzyme, and the effect of this is that they process alcohol much more slowly than others and become drunk on much less.

Molybdenum is also part of the metabolism by which algae dispose of sulfur, converting it to dimethyl sulfoxide and then, by means of a molybdenum enzyme, to dimethyl sulfide which is volatile. This gas rises from the sea into the atmosphere where it is oxidized to methane sulfonic acid which triggers cloud formation. It is also dimethyl sulfide which attracts sea birds to areas where the sea is rich in nutrients and is likely to have lots of fish.

World production of molybdenum is around 80 000 tons per year from mines in the USA, Australia, Italy, Norway and Bolivia, but current known reserves can only last about 50 years. Most molybdenum is converted to molybdenum sulfide which is used in oils as a lubricant and an anti-corrosion additive. Molybdenum is also used in catalysts, electrodes and molybdenum alloy steels, as support wires for filaments in light bulbs, and in X-ray machines.

Cobalt

The average person weighing 70 kilograms has around 3 mg of cobalt in their body, and we know this metal really is essential for humans because it is at the heart of the vitamin B_{12} molecule. Yet too much cobalt in the diet may affect the thyroid and damage the heart, and cobalt is a suspected carcinogen. The daily dietary intake is quite variable, and may be as much as 1 mg, but all except that in the form of the vitamin B_{12} will pass through the body unabsorbed. The amount we need is very small, and we probably have only about 1 mg in our whole body (less than 1/30 000th of an ounce). Foods rich in vitamin B_{12} are sardines, salmon, herring, liver, kidney, peanuts, peas, butter, bran and molasses.

The metal is commercially important and the main mining areas are Zaire, Morocco, Sweden and Canada. World production is 17 000 tons per year. Cobalt can be magnetised, like iron, and is used to make magnets. It also finds use in ceramics and paints. The radioactive isotope cobalt-60 is used in medical treatments and it can also be used to irradiate food. This preserves the food by destroying the organisms that cause decay, and the harmful bacteria that cause food poisoning.

Nickel

The average person weighing 70 kg has around 1 mg of nickel in their body. Nickel has been proved essential to some species, and it is related to growth, but its exact metabolism is not clearly understood. The human requirement could be as little as 5 mcg a day, but the daily intake is estimated to be around 150 mcg. Nickel occurs in the beans which contain the Jack bean urease enzyme, which contains 12 nickel atoms per molecule. Another relatively rich source of nickel is tea, which has 7.6 mg per kilogram of dried leaves. Other plants generally have less than half this amount.

Most nickel compounds are non-toxic, but some are poisonous, carcinogenic and teratogenic. Nickel carbonyl is extremely toxic. Some people are very sensitive to nickel, and since it is a component of stainless steel these people suffered nickel itch and eczema when stainless steel watch straps were introduced. The reason why nickel is thought to cause cancer is because it can substitute for zinc and magnesium, which are essential metal cations in DNA polymerase. The nickel ion is slightly different from zinc and magnesium ions, and so affects the behaviour of this catalyst, perhaps causing it to bind the wrong nucleotide, resulting in the formation of a wrong sequence of DNA. If it does this, and the mistake is not spotted and rectified, then we may have a cancerous cell. Happily for us the processes of checking, rectifying, inspecting and eliminating are very efficient and protect us every day from such a threat; but clearly if we are exposed to excess nickel, perhaps through our employment, then we are likely to increase the risk of its triggering a cancerous mistake.

World production of nickel metal is 510 000 tons per year and known reserves will last for about 140 years. The metal is mined in Russia, the USA, Canada and South Africa. Nickel is a silvery metal which is easy to work and can be drawn into wire. It resists corrosion even at high temperatures, and for this reason it is used in gas turbines and rocket engines. Nickel cadmium batteries can be recharged over a thousand times and still work efficiently, but the cadmium in these poses an environmental threat, as we shall see in Gallery 8.

Most nickel ends up alloyed with steel to give stainless steel, which is 74% iron, 18% chromium and 8% nickel. There are many other alloys of nickel, but the ones of the future are the so-called nickel superalloys used in rocket motors and jet engines, which experience temperatures of over 1,000 Celsius. The most useful is likely to be nickel aluminide, a remarkable alloy of nickel and aluminium, which was first developed at the US

Government's Oak Ridge National Laboratory at Tennessee. It is even likely to become part of the engine of the family motor car someday. Before then it will be incorporated into rockets, high performance jet engines and heat-exchangers. What makes nickel aluminide so special is its unique property at high temperatures. It is six times stronger than stainless steel, and it actually gets stronger the hotter it gets. At 800 Celsius it is twice as strong as it is at room temperature. The hotter you run an engine the more efficient it becomes, which is why the search is on for materials that will allow engines to run at red heat temperatures above 1000 Celsius. Nickel aluminide is one of the new breed of super-alloys that have sufficient strength to work at these temperatures.

GALLERY 3

STARTING LIVES, SAVING LIVES, SCREWING UP LIVES

An exhibition of molecules that can help and harm the young

▪ PROTECTING THE UNBORN ▪ MOTHER'S MILK ▪ SEXUAL CHEMISTRY ▪ THE HORNS OF A DILEMMA ▪ A KISS AT CHRISTMAS ▪ 'TWAS THE NIGHT BEFORE CHRISTMAS

GALLERY 3A: ▪ PRIVATE COLLECTION—RESTRICTED VIEWING ONLY

▪ ON THE WINGS OF A DOVE ▪ TURNING OFF THE DRUGS TAP ▪ NASTY HABIT ▪ SMOKE A CIG OR LICK A FROG ▪ PERCHANCE TO DREAM

IN THIS gallery we will look at the portraits of molecules which can affect us very profoundly, and not only ourselves, but also the life we carry inside us, or the life we would like to create. In a private room at the end of the gallery are a few portraits that were not thought suitable for public exhibition, but which selected individuals will be allowed to view. These are molecules that are deemed undesirable, but their eradication is proving difficult, if not impossible.

Few things are more important than creating new life, and yet nature has an almost cavalier attitude to the process, investing in gross overproduction of the raw materials necessary. Women have the ability to produce around three hundred eggs in a lifetime, and men to manufacture three hundred million sperm a week. Despite this abundance the human population has been kept in check in many ways—high infant mortality, famine, disease, war; but even so, today we have a world that is overpopulated with humans. This has come about through the success of science, which has lifted the first three of these natural scourges, although it has made the fourth much worse. Sadly science has so far not elicited the response of better birth control in many parts of the world, but it has made it possible to plan parenthood carefully. Science has also made it possible to ensure that if you decide to have a baby, then the baby you bring into the world should be perfect. The only prayer that potential parents in developed nations deem necessary is 'please let our baby be all right.' There are a few simple precautions that a woman can take to ensure her baby has a good chance of avoiding some risks that would seriously affect it. In this part of the gallery there are two molecules that she needs to think about.

Portrait 1

Protecting the unborn—folic acid

Folic acid is found in plants, animals and microorganisms such as fungi and yeasts. It is present in grass, butterflies' wings and fish scales. Humans need it also, as an essential component for several metabolic processes. And so does the foetus in the womb, and especially during its first few weeks of growth, because without folic acid it may not develop properly and could be born with the condition known as spina bifida. A baby with spina bifida has an exposed spinal cord, which may be so damaged that it causes paralysis of the legs. The condition can be diagnosed by looking for an abnormal protein in the amniotic fluid, or by ultrasound, and if it is detected then the mother can have the pregnancy terminated. Sadly not all such babies are detected, and some are born with spina bifida. However, thanks to modern surgery, they can still go on to lead fulfilling lives.

Spina bifida might be prevented if the mother-to-be has enough folic acid in her diet. Without folic acid, which the baby gets from her liver, there cannot be proper growth of the foetus. The woman can build up her folic acid store by eating foods such as liver, asparagus, spinach, lentils, navy beans (usually sold as canned baked beans), peanuts, mushrooms, yeast and bran, all of which have plenty. Some bread is now fortified with folic acid. Foods with the highest levels are liver, with 250 parts per million (ppm); brussel sprouts (100 ppm); spinach (90 ppm); broccoli (65 ppm); and oranges (50 ppm). Certain brands of cornflakes and bran cereals are now fortified with folic acid up to 250 ppm, but the richest sources are meat extracts, like Bovril, and yeast extracts, like Marmite and Vegemite, which have over 1000 ppm.

Folic acid has several effects on human metabolism, some so far unexplained, such as an increased tolerance of pain. It is needed for nucleic acid synthesis, growth and blood formation. A deficiency of folic acid leads to reduced antibody formation. Folic acid was once called vitamin M, although it is now recognized as one of the B group of vitamins. Its main function appears to be to help build other molecules, and its speciality is in supplying a single carbon unit. Folic acid is very good at picking up such carbons from other sources and then handing them over when needed to cell components, like DNA, and to amino acids, such as methionine.

We can store some folic acid in our liver, but when we are pregnant, old, or ill with diarrhoea this may become depleted if we are not getting enough from our diet—or from the bacteria which live in our intestines. They make folic acid too and this we can absorb, but they supply only a small part of our requirements. The level of folic acid in the body can be measured by analysing its concentration in red blood cells and plasma. Without enough folic acid we suffer a type of anaemia called megalobastic anaemia, and a quarter of pregnant women appear to suffer from this condition. For her baby's sake a woman needs to have enough folic acid soon after she had conceived—at a time when she may not yet have realized that she is pregnant. If she is lacking this vitamin then her foetus may suffer a neural tube defect, of which spina bifida is the most common; the neural tube forms about a month after conception. If a woman is planning to have a baby then health authorities advise she should take folic acid supplements, and the best way is to buy 400 mcg tablets from pharmacists or health food shops. One of these provides the recommended daily dose. Women of childbearing age can get the 400 mcg per day they need from their food, provided they know what to eat. Their liver stores folic acid and generally there will be a few weeks' supply in hand, unless the diet is deficient in foods which provide it.

Folic acid can be made synthetically from its molecular components, which are glutamic acid, p-aminobenzoic acid and pteridine. Pure folic acid will grow as deep yellow crystals from warm water, but it decomposes on heating in boiling water and it is also destroyed by light. For these reasons much of the vitamin is lost on cooking: those vegetables which contain a lot of folic acid are the ones we tend to cook the longest, such as brussels sprouts, cauliflower and spinach. Ideally these should be only lightly boiled.

Folic acid is defined as a vitamin for humans, in other words it is a vital component of our diet which we must obtain from the food we eat. Other living things, such as bacteria, can make all the folic acid they need, and this can be exploited to defeat them. The earliest antibiotics, the sulpha drugs, worked by blocking a key enzyme which bacteria use to make the folic acid they need, thereby preventing them from multiplying rapidly. This buys time for our body's antibodies to multiply and destroy them.

The role of folic acid in preventing neural tube defects was noted by Richard Smithells of the University of Leeds, England, in 1983, and as a result an extensive study was undertaken which involved 1800 women who previously had given birth to a spina bifida baby, and who were thinking of having another baby. They were all given various types of vitamin supplements, and half of them were given folic acid. A report published in 1991 showed that of the 1200 women who gave birth to a subsequent child, 27 had babies with neural tube defects. Of these, 21 were born to women not taking the extra folic acid, and only six were born to those women who had been asked to take folic acid.

Folic acid alone may not prevent all cases of neural tube defects, and this is the conclusion of a group of scientists from Trinity College, Dublin, Ireland. In 1993, in the *Quarterly Journal of Medicine*, they reported on a study of 56 000 pregnant women. According to one of the doctors in the group, Ann Molloy, the research showed that both folic acid and vitamin B_{12} were risk factors, and she argued that if folic acid is to be used to fortify food there may also be a case for including B_{12} as well. The Dublin team even suggested that recommended allowances of both vitamins were still too low. The World Health Organisation used to recommend a daily intake of 200 mcg of folic acid, but the trials with pregnant women used 500 mcg and today's guidelines suggest at least 400 mcg is needed. This is the amount every young woman should aim for.

Portrait 2

Mother's milk—arachidonic acid

Even if a foetus develops safely beyond the dangerous first three months, things can still go wrong and the body may reject the baby before it has come to term. Medical science and skillful nursing care can then come to its rescue, and yet technical assistance in the form of state-of-the-art incubators, instruments and monitors may not be enough.

Despite all the care that is lavished upon them, babies born prematurely are also slow to catch up, and are still smaller than normal babies, even when they are one year old. They also have a high risk of disabilities such as cerebral palsy and blindness. Research shows that they are probably disadvantaged by a lack of arachidonic acid in their diet. The placenta of a pregnant woman provides her foetus with plenty of arachidonic acid, and her breast milk continues the supply to her baby after she has given birth. A baby born prematurely is suddenly cut off from the arachidonic acid it needs, and it has to look for this from the formula feed that it is now given. Until recently most milk substitutes did not provide any arachidonic acid, and the level of this chemical in the blood of premature babies falls rapidly to less than a half that of a baby which is still in the womb.

Michael Crawford of the Institute of Brain Chemistry at Hackney Hospital, London, has been researching the effect of arachidonic acid on brain development for over 20 years. In 1992 he showed that a lack of arachidonic acid was a serious problem for premature babies, and that this was the main reason why they did not put on as much weight as expected in the weeks following birth. Crawford believes that arachidonic acid must be included in formula feeds so that these are as close as possible to the nutrition that the placenta provides. Research shows that premature babies collapse into a serious deficit of arachidonic acid immediately after they are born. Arachidonic acid, and the related fatty acid docosahexaenoic acid (often abbreviated to DHA), are essential for the growth and function of blood vessels within the brain, and for the brain itself.

The European Society for Paediatric Gastroenterology and Nutrition reported on the fat content of formula feeds in 1991. It recommended these should supply long-chain fatty acids like those in human milk, and in particular supply those polyunsaturated fats which a baby cannot make for itself. One group of these is known as the omega-6 fatty acids, in which the number 6 is related to the chemical structure, which consists of a long chain of carbon atoms with one or more double bonds at intervals along the chain. The 6 means that the first of these double bonds (also known as unsaturated bonds) is on the sixth carbon atom from the end of the chain, and it bonds the sixth carbon atom to the seventh. Arachidonic acid is an omega-6 acid.

Humans make the arachidonic acid they need from another, more common, omega-6 acid called linoleic acid, which we get from our food. Linoleic acid is especially abundant in seed oils such as sunflower oil (50%), in peanuts (14%) and in smoked bacon (5%), while the much maligned lard has 10%. Eventually a baby is able to do the same, once it has developed the necessary enzymes. Until then it has to rely on its mother to make the arachidonic acid it needs, but if it is born too soon then it has to be tube-fed using formula feed, or it may be fed its own mother's milk. (It was once common practice to feed premature babies on milk donated by other mothers, but this is no longer done because of the risk of HIV infection.)

Because the rate of development of brain and blood vessels is so great at this critical period, and arachidonic acid is so necessary, then clearly even formula feed for normal-term babies should also be fortified with this essential fatty acid. Unlike linoleic acid, arachidonic acid is rarely found in plants or animals, and its scarcity posed a problem for manufacturers of formula feed. The first to come up with an answer was the Dutch-owned company Milupa, which patented a process for extracting both arachidonic acid and DHA from egg yolks, and blending them into their formula in amounts and ratios similar to those of breast milk. Recently an American biotechnology company, Martek Biosciences, has found a way of extracting arachidonic acid from fungi (and DHA from algae), and other major formula feed companies have signed deals with them. Meanwhile, the debate continues over whether babies born normally need these fatty acids in their feed, because in theory they can make what they want from linoleic acid.

Arachidonic acid was discovered about 50 years ago and is present in liver, brain and various glands of the body. It is essential because we need it to make prostaglandins, hormones and cell membranes. More than half of the brain's structure is membrane, and this requires both arachidonic acid and DHA. Arachidonic acid may have other roles, and it has been shown that without linoleic acid and arachidonic acid rats develop eczema. Provided we get about 5 g (a fifth of an ounce) of acids like linoleic acid each day, we have enough to keep our skin in good condition.

Sometimes the cells of our body produce arachidonic acid when we would rather they didn't. A strained muscle, an infection, or an arthritic joint causes this to happen, and is the first step in the formation of prostaglandins, an excess of which causes inflammation and pain at the site of the damage. When this happens we can take painkillers such as aspirin, ibuprofen or paracetamol, and while these do not remove the arachidonic acid, they block the enzyme that turns it into prostaglandins.

Advances in food chemistry have resulted in a completely balanced formula for a baby's feed which provides all the nutrients it needs. A mother may have many reasons for bottle-feeding her baby, but if she breast-feeds then she may be providing a few extras as well, namely antibodies that she has made to protect against infections in the environment in which both she and her baby live. Breast milk can protect against life-threatening illnesses, such as gastroenteritis, which are more likely to afflict a bottle-fed baby. All babies are exposed to the dangers of bacteria, which can multiply in their stomach and intestines and cause vomiting, diarrhoea and dehydration, and whereas in developed countries a baby who becomes infected can be treated with antibiotics, in the developing world such an infection can cost a baby its life, and mothers there should always breast-feed.

The first milk a new mother produces, called colostrum, is particularly rich in the chemical lactoferrin, which is a natural antibiotic. Colostrum can have up to 15 g of lactoferrin per litre. As the days progress the amount of lactoferrin in breast milk decreases as the danger from germs recedes and the baby's own defences develop. Cows' milk has no lactoferrin because cows are milked long after they cease producing the bovine equivalent; however, this deficiency could be made good by adding lactoferrin to formula feeds. As yet, no brand has incorporated lactoferrin into its formula.

Lactoferrin works by trapping iron atoms, thereby preventing them from being taken up by bacteria which must have them if they are to multiply (see Gallery 2, page 42). We all have lactoferrin in our bodies, where it serves to move iron around and also to keep iron from generating free radicals, which it is apt to do. Lactoferrin is produced as required, and it is not something we need to take in with our diet. The parts of the body with particularly high levels of lactoferrin are tears, saliva and semen.

Portrait 3

Sexual chemistry—nitric oxide

In the spring a young man's fancy lightly turns to thoughts of love.

So wrote Alfred Lord Tennyson in his poem *Locksley Hall* in 1842. Today we know that young men think about sex at least four times an hour, and not only in spring. But whenever a young man's fancy may turn to thoughts of love, if those thoughts are to come to anything they must generate the molecule nitric oxide. Without this simple molecule nothing will happen to help him fulfil his romantic desires.

Nitric oxide is used by our body to relax muscles, to kill foreign cells and to reinforce memory. Nitric oxide relaxes the muscles of blood vessels and so will relieve an attack of angina in an older person. In our young man it will trigger an erection. Erotic thoughts and stimuli send a signal to the nerves of the corpus cavernosum, the spongy muscle in the penis, which then releases nitric oxide. This relaxes the muscle and so lets blood enter the tissues, causing them to swell. This role of nitric oxide was discovered in 1991 by a team headed by Karl-Erik Andersson of Lund University Hospital in Sweden, and it changed our perception of this curious molecule, which until 1987, was regarded only as a environmental pollutant from motor car exhausts, and a forerunner of acid rain. That it was part of human metabolism was unsuspected, and even when it was proposed that this might be the case, it seemed to fly in the face of its chemical nature because the molecule is a highly reactive free radical—it has a rogue electron. Such molecules generally survive for only a fraction of a second, but nitric oxide by itself is stable. However, put it with another molecule and chances are that it will react.

Nitric oxide is a gas that is easy to make in the laboratory—you simply add copper turnings to diluted nitric acid and collect the colourless gas over water. That way it will not interact with the oxygen of the air, which quickly converts it to brown fumes of the acidic gas nitrogen dioxide. Nitric oxide has been known for well over two hundred years, and it nearly killed the great chemist Sir Humphry Davy when he tried breathing it in 1800.

Nitric oxide consists of one nitrogen atom bonded to one oxygen atom. It is as simple as that. But what a wonderful molecule this is, not least for headline writers in the chemical press. Its chemical formula is NO and science magazines showered it with puns: NO sex; NO wonder; NO way; NO news is good news'; NO is the way our body says 'yes'.

When NO is released by cells on the inside of blood vessels, it relaxes nearby muscle cells and so lowers blood pressure. In 1987 Salvador Moncada and colleagues at the Wellcome Research Laboratories at Beckenham, England, were the first to realize that blood vessels could make NO, and a year later they discovered that it came from the amino acid arginine, of which there is a plentiful supply. Suddenly they were able to explain how a group of drugs, including amyl nitrite and nitroglycerine, work. These can stop a painful attack of angina by releasing NO, which relaxes the constricted vessels that are reducing the supply of blood and oxygen to the heart. You cannot treat people directly with NO because it is a toxic gas, as Davy discovered, but you can give them chemicals which release NO inside the body. It lasts there only for a few seconds, but this is long enough to do what is required.

Heart drugs have a long history. The effect on blood pressure of breathing amyl nitrite vapour was noted as long ago as 1867, and this is mentioned in the Sherlock Holmes story *The Case of the Resident Patient*. In World War I doctors reported that ammunition workers packing shells with the explosive nitroglycerine had very low blood pressures, and this led to the use of this compound as a vasodilator, one of a class of substances which will cause blood vessels to widen. (Like nitric oxide, both amyl nitrite and nitroglycerine are nitrogen oxide compounds.)

The second role of NO is to protect the body, which it does by killing unwanted cells. Microphages are cells in the blood that seek out foreign particles, such as invading bacteria or mutant cells, and destroy them by injecting a fatal dose of NO. Sometimes microphages are so active in the body's defence that they produce too much NO, even to the extent of being life-threatening. A leading cause of death among patients in intensive care is septic shock. As the body generates NO to fight the infection it may also lower the blood pressure to dangerous levels. Inhibitors which block the NO-forming enzymes can restore the blood pressure within minutes, and these are being used to treat such patients.

It is still not clear how a chemical like nitroglycerine, which we normally think of as an explosive, produces NO in the body. Scientists are now researching the way in which it operates, and eventually they should be able to produce a new range of longer-lasting NO-releasing drugs. Pharmaceutical companies hope to design compounds that are more effective in dealing with circulatory ailments than current treatments, and some are now being tested.

Nitric oxide also has a role in the body as a messenger. Because it is small it can diffuse into and out of cells easily, and it is quickly mopped up by oxygen. It may even be the looked-for 'retrograde messenger' that is the basis of memory. How does a receptor cell in our brain, which has once been stimulated, recognize the same stimulation again? The receptor does this by sending a 'message received and understood' signal to the cell that sent the message. NO serves as this messenger, and not only confirms that the message has been received but also programmes the sender cell to send an even stronger signal next time.

Nitric oxide was first identified in the brain by John Carthwaite and colleagues at Liverpool University, England, and at the same time Moncada's group at Wellcome discovered that the brain makes NO in just the same way as blood vessels do. These findings were confirmed when Solomon Snyder at Johns Hopkins University in the USA cloned the nitric oxide producing enzyme, NO-synthase, and found this was abundant in the brain.

The more we learn about NO, the more we can put other observations in context. We now discover that NO has been protecting our food for over a hundred years. Meat producers use sodium nitrite to inhibit dangerous bacteria from growing on cured ham and in tins of corned beef, although no one knew exactly why it worked so well. Now we understand: this simple salt works by acting as a source of NO. And even when we have eaten our corned beef sandwich NO may be there to help it on its way—because this is the molecule which triggers the wave-like contractions of the gut that moves food through our stomach and intestines.

Clearly the biochemistry of NO is still in its infancy and this simple molecule promises to provide us with many surprises in the future. The link with sexual arousal in men suggests that one day there could be an aphrodisiac chemical that would produce the desired effects by controlled release of NO at the appropriate site. Whichever young chemist finds the magic molecule, it should make his or her fortune. They might find greater satisfaction in knowing that it may also save the few remaining rhinos in the world, for reasons we shall now discover.

Portrait 4

The horns of a dilemma—keratin

Humans are fascinated by the idea of a simple love pill or potion that will arouse the object of one's desire, or wake the lazy member from its slumbers. The demand for aphrodisiacs has been constant down the ages, but almost all have proved worthless, although this has not stopped people buying them.

A large number of natural substances have been touted as successful aphrodisiacs. Some are the known sex attractants of other species, such as musk which triggers sex between deer and doe, or androstenone which does the same for pig and sow. These chemicals have no noticeable effects on humans although that has not hindered their use. Nor do any of the smells which the human body produces act this way, and so we turn to perfumes to make us attractive to other people. But perfumes are not aphrodisiacs, although they may put us in a romantic mood.

Ideally an aphrodisiac is something that we take that affects our emotions and sexual responses. Alcohol and marijuana can appear to be weak aphrodisiacs, but only because they make us less inhibited; and the Irish drink, Guinness, has an aphrodisiac reputation in some countries. Many foods are suspected of harbouring aphrodisiac chemicals, including bananas, asparagus, mushroom, mussels, figs, oysters, oats, sunflower seeds, nuts, avocados, carrots, celery, mangos, garlic and arsenic. Every culture has it favourite food that is supposed to boost our sexual drive and performance, but none is a true aphrodisiac.

The best we can say of these is that they may supply some essential vitamins and minerals in which the body may be deficient and which we know are needed to keep the reproductive system working properly. Zinc, which we examined in Gallery 2 (page 47), is essential for the functioning of the sex glands and the hormones that govern our sex-drive. Semen has a high level of zinc, and this continually needs to be replenished. A diet rich in this metal would clearly do a man no harm. Vitamin A is needed by the body to convert cholesterol into the male hormone testosterone, and if a man does not get enough vitamin A he would become sterile. However, he is rarely in danger of not taking in enough of this vitamin in a normal diet—indeed he is more likely to take in too much and provoke a toxic response.

There are molecules that have the ability to produce a stimulating effect on the genitals of both men and women: yohimbine and cantharides. Yohimbine is a crystalline compound that comes from the bark of the yohimbé tree (*Corynanthe yohimbé*) which grows in central Africa. Local men and women have used it for centuries to stimulate their sexual powers, and tests on rats and humans proves that the effect is real. As little as 10 mg of yohimbine is enough to trigger an erection in men and a tingling sensation in women, but too much is dangerous, and 3000 mg is a fatal dose. A man who injected himself with 1800 mg collapsed in a coma and only just survived. (The effect it had on him was, tactfully, not recorded.) In theory it should not be too difficult to find safer variants of this molecule, if we knew how it worked. Finding sponsors to fund the research might be a problem, although I suspect there would be no shortage of willing volunteers to test the new molecules.

The other aphrodisiac is cantharides which comes from the Spanish fly, the ground-up powder from the brilliant green beetle, *Lytta vesicatoria*, but there are risks because this chemical can kill. Spanish fly works due to its irritant effect along the walls of the urethra, the tube from which the bladder empties.

One molecule that has no aphrodisiac potency is the protein keratin, yet as rhino horn this has the reputation of being the most powerful aphrodisiac of all. Chinese medicine, which is currently enjoying a vogue in the West, prescribes rhino horn as a cure for such ailments as fever, arthritis, lumbago and male impotence. Perhaps not surprisingly many believe that it is therefore an aphrodisiac, which is why a couple of grams of rhino horn shavings costs about £20 ($30), making a complete horn worth several thousand pounds. It is not difficult to see why it is so highly prized. The horn symbolizes the bull rhino's prodigious sexual prowess: when he mates with a female, the sex act lasts an hour and during this time he can ejaculate a dozen times or more.

Of course rhino horn is not an aphrodisiac. It is just another form of the single-strand polypeptide known as keratin, which pigs grow as trotters, cows grow as hoofs, and humans grow as finger nails. Drinking rhino horn tea, the traditional way of taking it, will have the same effect as making tea from your own nail clippings.

Keratin is built up from all the common amino acids, but has an especially high content of one of them: cysteine. Because of this it has a large number of sulfur–sulfur cross-linking bonds between neighbouring chains, which gives it both physical strength and resistance to breakdown by protein-attacking enzymes. For these reasons it is useless as a food, let alone as an aphrodisiac. Nevertheless, rhinos continue to be slaughtered for their horns despite international agreements to outlaw the trade. In Africa there are fewer than 5000 rhinos left, and in India the numbers are thought to have fallen below 2000. If it becomes extinct, this will shame us all.

Yet rhino horn clearly meets a genuine human need for love potions, and as long as it is believed to have miraculous powers it will be used, and the fate of the rhino will hang in the balance. What can chemists do to help save these noble beasts? Very little, I suspect. We could make a synthetic form of rhino keratin which contained even more cysteine, and hint that it was stronger even than the natural material, but this would just be substituting one myth for another. The only way to stop the slaughter of the rhino is to educate people with a little chemistry so they know exactly what keratin is, and how useless it is as a medicine or an aphrodisiac.

Portrait 5

A kiss at Christmas—mistletoe

Kissing under the mistletoe was once part of Yuletide fun, an acceptable bit of innocent flirting between young people that society would condone for a season, but which it would normally frown upon. Although it became part of the Christmas festivities, the custom dates back to pagan times, when the plant was associated with fertility. Kissing under the mistletoe was supposed to lead to marriage and babies. Sadly, mistletoe has no magical properties in this respect, because it might help its survival if it had. There is a fear that it is becoming an endangered species in the UK, the country where the tradition originated in the rituals of the Old Religion of the Druids. The organization Plantlife is trying to revive the custom of hanging mistletoe in the home at Christmas, in the hope that this will create a market for it, and encourage people to grow it, thereby saving some of the ancient orchards where it is most likely to be found.

One reason for mistletoe's decline is the fear that mistletoe berries are poisonous, and while this is true, the danger is much exaggerated. The toxic chemical they contain has only recently been fully analysed, yet it is already in use as an anti-cancer agent in Germany.

Mistletoe is a parasitic plant which grows mainly on apple trees, poplars, willows and hawthorns. It is not entirely parasitic because mistletoe produces chlorophyll and this enables it to use sunlight to make some of its own food, although it draws water and other essential nutrients from the host trees. The main variety is *Viscum album*, which is one of around 1300 species worldwide. Some of them are even parasitic on other mistletoes.

It is rare to find mistletoe growing on an oak tree, the sacred tree of the Druids, but when it was it was treated with special reverence. The Roman author Pliny the Elder, writing in his popular science book *Natural History*, published in AD77, said that the Druids would cut the plant down with a golden sickle and catch it on a white mantle. It must never touch the ground as that would defile it. It was then used in human sacrifice, and proof of this has been found. In 1984 the well-preserved naked body of a young man was found in the peat bog at Lindow Moss, Cheshire, England, and he had eaten mistletoe before he was ritually murdered. The sacrifice was dated to around 300BC.

There are two main families of mistletoe, distinguished by their white or red berries, of which the white ones are toxic. The berries are rich in glucose, and in some parts of the world, such as South Africa, mistletoe has been fed to cattle as fodder in times of drought, and even eaten by humans. The berries are filled with a semi-transparent pulp which has great sticking power. Crushed berries were the main component of the infamous bird lime, which was smeared on twigs to catch small birds.

The white berries are not life-threatening, but eating them causes stomach cramps and diarrhoea. *Poisonous Plants and Fungi*, a guide published by the UK Ministry of Agriculture, Fisheries and Food, recommends induced vomiting as first aid for anyone who has eaten them. Despite its toxicity, earlier generations used mistletoe as folk medicine. The juice from the berries was smeared as an ointment to relieve strains, sores, dandruff, warts, ringworm and impetigo, which is a contagious bacterial infection of the skin. An infusion of the berries was drunk as a medicine for epilepsy, colds, fevers, syphilis, gout and worms. It was considered particularly potent against infertility in humans and cattle, and for easing the pains of childbirth. Of course it was useless against all these ailments, but its psychological benefits probably compensated for the lack of curative properties.

Mistletoe is not totally useless. Research earlier this century showed that mistletoe extract acts as a diuretic and reduces blood pressure, and is an antispasmodic agent. The active ingredient in the berries is a substance known as mistletoe lectin. At the Madaus pharmaceutical company in Cologne, Germany, the head of research, Hans Lentzen, is directing a project to develop mistletoe lectin, which is already being given to cancer patients undergoing chemotherapy and radiation treatment. According to Uwe Pfüller of the medical faculty of the University of Witten-Herdecke, the main effects are a higher quality of life and its prolongation, and there is tumour regression. He believes that mistletoe lectin not only kills cancer cells, but also stimulates the immune system.

In 1995 the structure of mistletoe lectin was finally solved by Rex Palmer and Edel Sweeney at Birkbeck College, London, after six years of work. They achieved their results by the analysis of lectin crystals using X-rays generated by electrons travelling close to the speed of light. They discovered that the toxin consists of two pairs of large proteins, each pair made up of one part which can attach itself to cells walls, and an enzyme part which wreaks havoc on the target cell by preventing essential proteins from being made. Currently ways are being sought to improve the effectiveness of mistletoe lectin by attaching the toxic enzyme to an antibody which seeks out cancer cells that it can then destroy. It might also be used to control white blood cells to prevent the rejection of organ transplants. Palmer is now able genetically to engineer mistletoe lectin, and he plans to modify its two parts to make the binding part more effective, and in that way to target cancer cells.

It looks as though mistletoe could have something to offer humans after all, but not as originally conceived. The midwinter tease of standing invitingly under the mistletoe, waiting to kiss or be kissed, may no longer be part of the Christmas festivities, but as it disappears so will the mistletoe, and with it a tiny part of our pagan heritage. However, if it brings relief to some cancer sufferers, then we might well feel inclined to continue the cultivation of this remarkable plant.

Portrait 6

'Twas the night before Christmas—penicillin

Penicillin, the chemical that has brought relief to hundreds of millions, also has a strange story with a Christmas theme. It was on 24/25 December 1940 that its unsung hero, Norman Heatley, completed preparations to make penicillin on a scale that was big enough to treat a human patient for the first time.

Alexander Fleming discovered penicillin by accident in 1928. A rogue spore of the mould *Penicillium* had landed on one of his culture dishes. As Fleming reported: 'It was astonishing that for some considerable distance around the mould the staphylococcal colonies were undergoing lysis.' In other words something was dissolving the deadly microbes and killing them. Fleming cultured some more of the remarkable mould and found that, even when he diluted the broth in which it was made to a hundredth of its strength, it still killed bacteria such as staphylococci, pneumococci and streptococci. Fleming sent samples of the mould to other laboratories, but no one was able to extract the chemical that was acting as a powerful antibiotic. It seemed destined to remain merely of academic interest. By the outbreak of the World War II, in September 1939, little progress had been made, yet by D-Day on 6 June 1944 the Allies had in stock enough penicillin to treat all those who were wounded and needed it. Total penicillin production was 5 billion units in 1943, but by the end of 1944 it had reached 300 billion units, enough to treat 500,000 people a month. Within ten years of the end of that war penicillin was saving millions of lives all over the world as one bacterial disease after another fell before its onslaught.

There is a dark side to the penicillin portrait: instead of being a scientific triumph for the people of the British Empire, penicillin was a poor investment. Although it was discovered, researched and first produced in Britain, its citizens had to pay royalties to US companies whenever they used it. How this strange state of affairs came about we will discover, but first let us look a little closer at a brighter, though often neglected, part of the painting. Standing in the background is the figure of Norman Heatley. Without him penicillin might never have been a commercial success.

Heatley was born in the small town of Woodbridge in the county of Suffolk, England, where his father was a vet. Heatley went to Cambridge to study science, and after he graduated in 1933 he did a PhD in biochemistry. From there he went to a temporary job at the Sir William Dunn School of Pathology, part of Oxford University. There his practical skills were to prove invaluable, and it was largely thanks to the technical ingenuity of Heatley that enough penicillin was produced for animal tests.

The breakthrough with penicillin came on Saturday 25 May 1940, when Heatley watched hour-by-hour over eight mice, each of which had been injected that morning with 110 million streptococci, a virulent strain which would kill them within a day. Four of them were injected an hour later with doses of Heatley's penicillin solution, the other four remained as controls. By late afternoon the four controls were already very sick, and they began to die soon after midnight. By 3.30a.m. all were dead. The four treated with penicillin were fine. Heatley cycled back to his rooms through wartime, blacked-out Oxford, to snatch a few hours sleep before returning to the lab to tell the good news to his supervisor, Professor Howard Florey, head of the Dunn School. 'Looks quite promising', was Florey's grudging assessment of Heatley's report. In fact, it was little short of a miracle.

Perhaps Florey was right to be cautious, because he knew how difficult it would be to scale up production of penicillin so that they would have enough to treat a human being, who, being about 3000 times heavier than a mouse, would need a correspondingly large dose of penicillin. The makeshift pie dishes, biscuit tins and hospital bed pans that Heatley had been using as fermentation vessels were no longer adequate. Pharmaceutical companies, who had the equipment to culture micro-organisms in the amount that Florey required, were approached but were unwilling to divert people and resources from the war effort into producing what was still an untested drug. The Battle of Britain was about to begin and for the next few months, until June 1941, the Nazi Luftwaffe would bomb towns and cities, night after night. Thankfully, Oxford was spared.

Heatley's only recourse was to try and produce penicillin in vast amounts at the Dunn School itself. He had to find a way to grow the *Penicillium notatum* mould, from which the active agent was extracted, on a much larger scale. Heatley himself had devised the extraction method which had enabled the first animal tests to go ahead, and which was to be used eventually on a commercial scale. Now to grow more *Penicillium*, special culture flasks were designed. These were the size of a large book, with a spout at one corner, and were ideally shaped for stacking in large numbers for sterilization in the department's small autoclaves. But enquiries at the Pyrex glass company frustrated the plan: the vessels would be too costly to make and take six months to deliver.

Then Heatley came up with a brilliant idea: the penicillin need not be grown in glass vessels—it might be just as easy to culture in porcelain vessels. These could be made much more cheaply, and if they were glazed only on the inside they would remain rough on the outside, making them easy to handle. (Heatley still keeps one of these historic vessels to show to interested visitors.)

Florey was persuaded to try Heatley's plan, and he wrote to Dr J.P.

Stock of Stoke-on-Trent, sending him drawings of the type of vessel they required and seeking his help. Stoke-on-Trent was renowned for the manufacture of fine pottery for hundreds of years, and was where Josiah Wedgwood lived and worked (1730–1795). Stock contacted the firm of James MacIntyre and Co., who said they were willing to make the vessels.

Heatley caught the train from Oxford to Stoke-on-Trent, but his 100 mile journey was to take a whole day because of an air raid on Birmingham through which they had to pass. However, on his arrival at the MacIntyre factory he was amazed to find that they had already made model vessels for him, and with a few minor adjustments they were ready for firing and glazing, a process which took about three weeks.

In late November 1940 three trial vessels arrived at the Dunn School and proved completely satisfactory. An order for several hundred was placed and on 23 December Heatley borrowed a van and brought the first batch of 174 back to Oxford. On Christmas Eve 1940 Heatley and his fellow workers spent the day washing, sterilizing and filling dozens of the new fermentation vessels which he had designed. On Christmas Day Heatley returned to the lab and seeded them with the spores of the fungus *Penicillium notatum*. He then stacked the vessels for their ten-day period of incubation, at the end of which time he hoped the liquid on which the fungus was growing would contain enough penicillin to begin tests on humans.

A month later Heatley had extracted partly purified penicillin from 80 litres (about 20 gallons) of crude penicillin solution. With around 1–2 units of penicillin per millilitre (ml) this amounted to around 100 000 units in all—a unit being defined in terms of potency measured on a special assay culture plate invented by Heatley. (A few years later, it was shown that a unit was equivalent to 0.6 mcg of pure penicillin, but by then developments and commercial production had increased the yield of penicillin to 40 000 units per ml.)

Florey now thought they had enough penicillin to try it on a human patient. In the Radcliffe Infirmary in Oxford lay a 43-year-old policeman, Albert Alexander, dying of staphylococcal and streptococcal infection contracted when he had scratched his face on a rose bush a few months earlier. Despite the efforts of doctors, he was covered with suppurating abscesses all over his head, one of which had necessitated removal of an eye. Nothing had helped, not even sulphonamides—these are not effective when a patient is saturated with pus. On 12 February he was given an infusion of penicillin and immediately started to improve. More penicillin followed, and more improvement. On 19 February he was well on the way to recovery and treatment was stopped. His condition was stable for about ten days, then it began to deteriorate, and he died on 15 March. By then the supply of penicillin had been used up on another patient.

This was Kenneth Jones, a lad of 15 who had had a hip operation on 24 January 1941 to insert a pin, but whose wound had become septic and who again had not responded to sulphonamides. His temperature had been over 100°F for two weeks, when, on 22 February, he was given penicillin, some of which had been extracted from the urine of the now recovering Albert Alexander. Within two days the boy's temperature was normal and four weeks later he was fit enough to have another operation to remove the pin that caused the infection. Kenneth Jones recovered completely and died in 1996.

Heatley set about making more penicillin, and by May 1941 he had enough to enable other patients to be treated successfully. Sometimes the success was bitter-sweet: one sad case was John Cox, aged 4, who had cavernous-sinus thrombosis from septic spots on his face which had followed measles a few weeks before. Cavernous-sinus thrombosis was a condition that was invariably fatal. He responded to penicillin treatment and was almost better when he suffered a ruptured mycotic aneurysm and died. The autopsy revealed that the infection of the cavernous sinus had been arrested.

The results of these and several other cases were described in a landmark paper modestly entitled 'Further Observations on Penicillin', which was published in the *Lancet* (16 August 1941, pages 177–201) under the names of E.P. Abraham, E. Chain, C.M. Fletcher, W.H. Florey, A.D. Gardner, N.G. Heatley and M.A. Jennings.

There was no doubt that the Dunn School had proved the efficacy of the new drug. What was now needed was to speed up production. The Rockefeller Foundation, which had been supporting Florey, urged him to visit the USA and to seek help from firms there. In July 1941, Florey and Heatley flew to New York. It was at the Agriculture Department's Northern Regional Research Laboratory at Peoria, Illinois, that their pleas for help fell on fertile ground. The head of the fermentation division, Dr Robert Coghill, agreed to undertake a large-scale project using the mould brought by Heatley from Oxford. Heatley stayed on at Peoria, while Florey began a series of visits to try and interest US drug companies, apparently without success, although some now began their own experiments on penicillin production.

Meanwhile, Heatley was assigned to work with Dr Andrew J. Moyer, who suggested adding corn-steep liquor, a by-product of starch-extraction, to the growth medium. With this and other subtle changes, such as using lactose instead of glucose, they were able to push up yields of penicillin to 20 units per ml. But their cooperation was becoming rather one-sided, and Heatley wondered why Moyer had become very secretive, no longer confiding in his British colleague. In July 1942 Heatley returned to Oxford, and was soon to learn why. When Moyer published their research results he omitted Heatley's name from the paper, despite an original contract which stipulated that any publications should be jointly authored. Later he was to learn that Moyer had a good reason for taking all the credit for himself. To have acknowledged Heatley's part in the work would have made it difficult to apply for patents with himself as sole inventor, which is what he did.

Meanwhile Heatley had other things on his mind—his future employment. His contract with the Dunn School was coming to an end, and Heatley, thinking it would not be renewed, applied for another job with a chemical firm and had been accepted. When Florey discovered what was happening he berated Heatley for wanting to leave. In the end Heatley stayed on for the rest of his career, researching antibiotics and writing or co-authoring 65 scientific papers.

Penicillin's success brought Florey great acclaim: he was knighted in 1944, became Baron Florey in 1965, and received the Order of Merit, Britain's highest accolade for creativity. With Ernst Chain, who also worked at the Dunn School, and Alexander Fleming, he shared the 1945 Nobel Prize for Physiology and Medicine. Fleming and Chain were also to be rewarded with knighthoods by a grateful British Government.

Our hero, Heatley, was not quite overlooked. When he retired 35 years later in 1978 he was rewarded with an OBE (Order of the British Empire), and in 1990, when Heatley was almost 80, the University of Oxford granted him the honorary degree of DM. But Heatley was to outlive the others, and at the time of writing he lives in the same modest house he and his wife, Mercy, bought in 1948 in the picture-postcard village of Old Marston, a few miles north of Oxford.

The University of Oxford was never to receive any of the financial rewards that penicillin generated around the world in the next twenty years. All these flowed to US companies and individuals. The reason was that Florey had been persuaded by Sir Henry Dale, head of the Medical Research Council, that it was unethical to patent a medical discovery, so he didn't. The result of this foolish advice was that for the next 25 years the British also had to pay royalties on the wonder drug that had been discovered, researched and developed in Britain.

Never in the field of human conflict against disease did so many lose so much for no reason.

GALLERY 3a

STRICTLY PRIVATE

Please note that admission to this part of the gallery is restricted

Portrait 7

On the wings of a dove—ecstasy

In his novel *Brave New World* Aldous Huxley pokes gentle fun at a future world run according to the dictates of science, genetics and determinism. It is a stress-free world, due mainly to its inhabitants' use of 'soma', the perfect anti-depressant pill that is freely available to all.

In the real world of today, those suffering from clinical depression can be prescribed various pills, such as Valium or Prozac by their doctors but for the rest of us, who need a little relief from the stresses of life, alcohol and nicotine are legitimately available. There are also several illegal substances that offer temporary escape, but most of these are regarded as dangerous. One type, the amphetamines, appear safer than most, and some people are in favour of these being used as recreational drugs. Indeed, when Huxley wrote *Brave New World* in 1932 one of the amphetamines was already being used by doctors—as a nasal decongestant. An even safer variety had also been made, but was not used. Today we know it by the name of ecstasy. Yet even ecstasy can kill.

Ecstasy was originally intended to be an aid to dieting. It was patented in Germany in 1914 by the pharmaceutical company E. Merck as a treatment for obesity, but it was never marketed. Today it is big business, with an estimated 100 million tablets a year illegally sold in the UK alone. Its chemical name is 3,4-methylenedioxymethamphetamine (abbreviated to MDMA), but it is as 'ecstasy' or 'E' that it is best known. It is listed as a forbidden substance under Schedule 1 of the Misuse of Drugs Act.

Ecstasy works by changing the amounts of various chemicals in the brain. It triggers the release of dopamine, which makes us feel good, and noradrenaline, which both makes us energetic and counters the action of serotonin which governs sleep—so we feel wide awake. This explains why it is popular with all-night ravers, who feel nothing worse than a mild hangover the following day. Nor is it addictive, although it may have some mild hallucinogenic effects.

When ecstasy is taken, it acts on the nerve cells at the base of the brain causing a pleasurable high. Prolonged taking of ecstasy can damage the axons which link these cells to other parts of the brain, although they will regrow. According to users, the effect of ecstasy lasts about two hours, but it can be prolonged by taking a tablet of Prozac at the same time, and this makes the 'come down' easier so there is no hangover.

The current craze for ecstasy can be traced to Californian psychotherapists in the 1960s who started treating their clients with MDMA, supposedly to help them gain confidence in dealing with other people. California law allowed MDMA to be used under supervision, but it soon appeared on the streets, and clandestine laboratories started making and distributing it. By the mid-1970s, ecstasy was in widespread use across the USA. Warnings about it began to appear in the early 1980s, following tests that showed it produced brain cell changes in rats. In 1996 a group of doctors reported in the *British Medical Journal* that even a single dose of ecstasy produces long-term damage to brain cells in monkeys. The US Food and Drugs Administration classified it as a Schedule 1 drug under the US Controlled Substances Act in 1985, although the psychiatrists who used it opposed this move. Others thought the FDA was overreacting because ecstasy seemed to be relatively free of harmful side-effects compared to other illegal drugs then in use.

John Davies, Professor of Psychology at Strathclyde University, Scotland, and author of *The Myth of Addiction*, sees the current campaign against ecstasy as misdirected: he points out that even using the most alarming estimates, ecstasy is not a major cause of death among young people; and that indications are that at most there is only one death per two million tablets taken. By comparison, Davies claims, accidental deaths due to adverse reactions to prescribed penicillin are about ten times as high. Davies believes that control and regulation are a better way forward, and that we have to learn to live with substances like ecstasy, and to concentrate on reducing the harm they do. He suggests that a war on drugs, and the unattainable aim of stamping them out, is misconceived.

Amphetamine is the basic molecule from which others are derived. It is a simple chemical derivative of benzene, to which is attached a short chain of three carbon atoms with a nitrogen on the middle carbon. Converting it to methamphetamine (also known as 'speed' or 'ice') simply requires an extra carbon on this nitrogen. Amphetamine itself was first prepared in 1897, its stimulant effect was only discovered in 1928. It was then marketed as Benzedrine, and prescribed as a nasal decongestant because it acted to reduce swollen mucous membranes. However, its stimulatory effects were more striking, and both it and the chemically modified form methamphetamine (Methedrine) were widely used by armed forces in World War II. Bomber crews took it to keep them awake during long flights, and soldiers were given it to counteract battle fatigue.

Surplus supplies of these drugs were sold openly after the war, especially in Japan, and by the mid 1950s as many as two million people were taking them as a daily stimulant. In Britain amphetamines were traded in the discos of the 1960s as 'purple hearts', a mixture of amphetamine and barbiturate. Doctors still prescribe amphetamines for narcolepsy, which is a rare condition characterized by an irresistible desire to fall asleep during the day.

Both amphetamine and methamphetamine act to stimulate the central nervous system, and to increase blood pressure and heart rate. Methamphetamine is preferred for medical treatment because it has more of a stimulant effect while having less effect on blood pressure. MDMA also behaves in the same way as these amphetamines, but its effects are more pleasurable because this molecule also fits the brain receptors which control serotonin and release dopamine. The MDMA molecule is like methamphetamine, but it has a five-membered ring joined to the benzene ring, sharing two atoms. This extra ring consist of two oxygens linked through a carbon. MDMA is an oil, but it is easily converted to solid by reacting it with hydrochloric acid. This forms a chloride salt, the white powder known as ecstasy.

Ecstasy first appeared in British nightclubs in 1989, and the National Poisons Unit estimates that it causes about 20 deaths a year. The few victims who die from ecstasy die of heatstroke, because the drug raises the body temperature by about 4 Celsius, which in the hot atmosphere of a rave could reach the critical 5 Celsius at which irreversible changes may occur and vital organs fail.

Portrait 8

Turning off the drugs tap—cocaine, heroin and designer drugs

It is possible, at least in theory, to suppress the illicit drugs, cocaine and heroin, by controlling the materials needed for their manufacture. This idea is behind the lists of proscribed chemicals drawn up by the Home Office in the UK and the Drug Enforcement Administration (DEA) in the USA. A list of 22 restricted chemicals to be controlled was approved at international level in an agreement signed at the Vienna Convention Against Drug Traffic of 1988. Some countries, go even further than this: the USA, for example, has a control list of 33 chemicals.

Two categories of chemical are identified. There are the *essential* chemicals, and these are needed to extract natural plant materials, such as heroin from the opium poppy or cocaine from coca leaves. Essential chemicals are reagents such as acetic anhydride and potassium permanganate, or solvents such as methyl ethyl ketone and diethyl ether. Purchasers of these have now to provide a valid reason for needing them if they want more than a minimum amount. This strategy is not proving very effective because, while it can have a limited effect on cocaine production, it can have almost no effect at all on heroin production.

The other category is precursor chemicals: those which eventually end up as part of the drug molecule itself. With the willing cooperation of the chemical companies, controlling these chemicals could eventually eliminate the illegal manufacture of speed (methamphetamine), angel dust and other designer drugs. Angel dust is phencyclidine or PCP. This compound was developed in the 1950s as a general anaesthetic, but a third of patients experienced pleasant hallucinations as they came round from their operations.

Heroin is impossible to control by indirect means. This substance comes from opium, which contains about ten per cent morphine. This is first extracted from the poppy heads, and then reacted with acetic anhydride to convert it to heroin. Over a million tons of acetic anhydride are produced throughout the world each year, and used to manufacture a large range of products from low temperature detergents to printing inks. It would be almost impossible to prevent heroin producers getting their hands on the small amount they require. Nevertheless, acetic anhydride is now listed as an essential chemical by the DEA, and all sales over a thousand litres must be notified.

Cocaine is a different matter, and it should be possible to stop its production because it requires a few essential chemicals to extract it from coca leaves and refine it. Cocaine production in Columbia is probably in excess of 1000 tons a year, and this requires the importation of 20 000 tons of organic solvents, which is 15 000 tons more than legitimately required by industry. Along the way to the final snowy white powder, the process requires the solvent methyl ethyl ketone, and the oxidizing agent potassium permanganate. Colombia imports over a third of all the methyl ethyl ketone shipped from the USA to South America. Clearly much of this is destined for the cocaine laboratories. The Colombian authorities are also trying to control diethyl ether, which can be used in place of methyl ethyl ketone, and have not issued an import licence for this solvent since 1987. However, the police seize almost 1,250,000 litres (250 000 gallons) of it a year when they raid cocaine labs, and estimate that each year ten times this amount is imported into the country illegally.

The chemical industry which manufactures these solvents is very aware of the problem and keeps a strict check on bulk sales, but cannot control individual purchasers along the distribution chain as the solvent moves from tanker to drum. A drum of solvent can now fetch many times its normal price on the Colombian black market. Clearly the control of chemicals does hinder the cocaine industry, even if it does not stop production. A lot of Colombian cocaine is now exported to Europe, and seizures there have risen to over 20 tons a year with single loads of over a ton sometimes being discovered.

For designer drugs, control is possible by cutting off the supplies of precursor chemicals. For example, angel dust's phencyclidine (PCP), which is cheap and easy to make, was set to become the most popular street drug in the USA prior to the appearance of crack, the smokable form of cocaine. Today angel dust is on its way out and will never regain its hold, as piperidine, the chemical needed to make it, is now a DEA listed chemical. Likewise speed needs ephedrine or phenylacetone, both of which are controlled, and this street drug is also relatively rare. In theory ecstasy should by now have been eliminated, but the control of its precursor chemicals has been less successful.

The rewarding effects of amphetamine and cocaine are produced by interfering with the pathway between the 5-hydroxytryptamine (5-HT) neurons of the midbrain and the higher processing centres in the forebrain. This pathway relies on dopamine as the chemical messenger, and changes to the release of this produce the pleasurable effects of these drugs. This same system may also play a role in the rewarding effects of other drugs of abuse such as heroin, nicotine and alcohol.

Portrait 9

Nasty habit—nicotine

Nicotine is a drug that is both intellectually stimulating and emotionally relaxing. It is legal, readily available, painless to administer, and doesn't cause cancer. But there is a snag. To get our fix we need to smoke tobacco, and tobacco smoke is dangerous. There are other ways of taking in nicotine but these are promoted not as ways of taking the drug, but rather as ways of stopping taking the drug. One new way of dosing up on nicotine is through the skin: waterproof patches which are impregnated with the drug release a measured amount to be absorbed by the body. These patches are expensive, but they are less expensive than smoking 20 cigarettes a day, and in the long run may even save your life. As yet there is no evidence that nicotine patches alone have produced nicotine addiction.

Nicotine comes from *Nicotiana tabacum*, the tobacco plant, which gets its name from Jean Nicot, the French ambassador to Portugal, who sent some seeds to Paris in 1550. It is a distant cousin of nicotinic acid, better known as the vitamin niacin, and this is produced commercially from nicotine by treating it with nitric acid. But unlike the vitamin, nicotine can be a deadly poison, and nicotine sulphate was once commonly used as a powerful insecticide.

Smoking is the quickest way to get your fix of nicotine—one puff, inhaled, acts within seven seconds. It passes from the lungs into the arteries and goes straight to the brain, but its effects are short-lived. Research shows that although the half-life of nicotine in the body is about two hours, its level in the brain falls more quickly, which explains why regular smokers need a cigarette about once every hour. The same levels of nicotine in the blood and the same degree of nicotine dependence is found among those who use dry nasal snuff, which was a popular way of taking nicotine in the eighteenth and early nineteenth centuries, or who chew tobacco, which was popular in the nineteenth and early twentieth centuries, or who keep a pouch of moist oral snuff in the mouth, a method of taking nicotine which began in the twentieth century. While we sleep, the nicotine level in our body falls and we lose part of our tolerance to it. This is why the first cigarette of the day is regarded by many people as the best: its effects are the most dramatic.

Until about 1980 nicotine was not considered a drug of addiction. It was thought to be an acquired habit, like drinking tea, and one that should be easy to break—in other words if we decide not to smoke we simply stop doing so. It has since become accepted that nicotine is that far from being merely a pleasant diversion. It is far more compelling than that: the desire for nicotine results in an intense craving in smokers who go without cigarettes even for just a few hours. If nicotine taking is not quite an addiction, it is certainly a habit that is difficult to break.

Even so, nicotine has its attractions: it cheers us up, improves our concentration, increases our ability to learn, counteracts our craving for sweet foods (so we are less likely to be overweight) and reduces stress. Cigarette adverts used to list these as the benefits of smoking. The claims stem from the main effect of nicotine, which is to increase the secretion of dopamine. This extra boost of dopamine reduces feelings of anxiety in those under stress. But there is a price to pay: repeated doses of nicotine gradually increase the number of nicotine receptors in the brain, which is why smokers experience such intense cravings for many weeks when they try to kick the habit.

Though nicotine is taken daily by hundred of millions of people around the world, it is classed as a highly toxic substance. Irving Sax in *Dangerous Properties of Industrial Materials* lists nicotine as a colourless, odourless, sharp-tasting liquid which causes vomiting, diarrhoea and convulsions if we ingest it, breathe it in, or absorb it through the skin. A fatal dose for the average adult is 60 mg (two drops of pure nicotine). We may experience symptoms of poisoning the first time we smoke a cigarette and absorb about 1 mg.

So how much nicotine does a smoker need to satisfy his or her craving? A daily packet of 20 cigarettes will contain up to 30 mg—that is about half the fatal dose. Not all of this is available to the smoker, but most is. Nicotine patches contain 50 mg, of which about 20 mg is absorbed by the wearer. Controlling the rate at which this absorption happens, and at the same time preventing the loss of nicotine, requires a special kind of patch with some sophisticated technology. To prevent the nicotine evaporating, or being leached out if the wearer bathes, the patches have a top layer of polyester, the plastic used for carbonated drinks bottles, with an inner film of aluminium. Below this is a layer of rayon impregnated with the nicotine. The release of this to the skin is controlled by a lining of polyester and three layers of plastic adhesives. The patches are supplied in child-proof packaging.

As people adjust to taking their nicotine through the skin, the craving for cigarettes should fade. As it does, they wear patches with less and less nicotine until they are cured. The effectiveness of the patches has been confirmed in a series of double blind tests. Smokers wishing to stop, and who used the patches, were twice as likely not to be smoking after three months as those who were given a placebo patch. Also those on nicotine patches did not gain weight, whereas those on the placebo put on as much as ten pounds.

Portrait 10

Smoke a cig or lick a frog—epibatidine

At first glance, the scientific paper entitled 'Epibatidine: a novel (chloropyridyl)-azabicycloheptane with potent analgesic activity from an

Ecuadorean poison frog' looks mildly interesting. Its authors are Thomas Spande, Hugo Carraffo, Michael Edwards, Herman Yeh, Lewis Pannell and John Daly of the US National Institute of Diabetes and Digestive and Kidney Diseases, National Institutes of Health, Bethesda, Maryland. They had extracted and identified the toxic chemical with which a tropical frog arms itself against its predators. But they published their results in the world's leading chemistry journal, the *Journal of the American Chemical Society* (1992, volume 114, page 3475), which suggests that there is something of special interest to chemists in the paper. There is.

Not only is epibatidine very toxic, and the reason why this frog was used by native Indians to make poison arrows, but it also turns out to be a superb painkiller. This painkilling ability panders to the popular belief that nature somewhere has a cure for every human affliction. Epibatidine is also an organochlorine compound, which confounds somewhat the environmentalists' belief that organochlorines are entirely manufactured chemicals that cause disease and damage the environment. This belief comes from years of campaigning against damaging organochlorines such as DDT, dioxins and CFCs. Epibatidine too is highly dangerous, but it is perfectly natural. It would seem a little unfair on the frogs to eradicate them because they are making a dangerous organochlorine molecule.

Sadly, the researchers had had to kill 750 of these beautiful red-and-white striped creatures to get enough material to analyse, because each frog yielded less than a tenth of a milligram of the compound for analysis. This was a one-off sacrifice, because they discovered it was a simple molecule, easily made in the lab. They also showed it was two hundred times stronger than morphine, and this they were able to demonstrate using the hot plate test. In this rather primitive test a rat is given the painkiller and dropped on to an electrically heated plate. Normally the rat would immediately leap into the air, but those which had been given a dose of only 5 mcg of epibatidine just stood there and waited, unaware of the damage the heat was causing.

Even more puzzling was the observation that when a rat that had been given epibatidine was injected with naloxone, the chemical which neutralizes painkillers like morphine, it still failed to respond to the searing heat. Clearly this new analgesic did not work by blocking the usual pain receptors. Had the animal's brain some other mechanism for controlling pain? In fact it turned out that epibatidine was targeting the nicotine receptors. Nicotine and epibatidine molecules have very similar shapes, the difference being that epibatidine binds more strongly. The obvious deduction is that this molecule is even more potent than nicotine and might have the same effects.

As we saw with nicotine, there are some benefits as a result of the higher levels of dopamine this produces in the brain, but the price we pay for using it is high. Could epibatidine offer a safer alternative to nicotine? Or would it be just too risky?

Epibatidine tantalizes chemists for other reasons. The brightly coloured tree frogs find it an effective deterrent to predators and it is highly toxic, which is no doubt why they confine it to their skin. So how do they produce it? Reared in the comfort and safety of captivity they fail to make any epibatidine at all, suggesting that there must be something in their natural diet required to make it, or perhaps that they only produce it when living in the wild under constant threat from predators.

We have seen that the compounds that nature has provided, and which we misuse for our enjoyment, are really rather crude, and that with a little fine tuning by chemists they can be made safe. This happened with the natural painkiller salicylic acid, which was originally used to treat fevers, despite its corrosive effects on the stomach. Today it is taken by millions as in the much-safer, chemically-modified version we know as aspirin. The epibatidine story is only just starting. It might well end with a better painkiller, or pill that smokers can take if they want to stop smoking. It might even result in a pill that will enhance learning or improve our enjoyment of other intellectual pursuits.

Portrait 11

Perchance to dream—melatonin

Methought I heard a voice say, 'Sleep no more! . . .'

Shakespeare, *Macbeth*, Act 2, scene 2

Many executives are overworked and under stress, and a characteristic of such lifestyle pressure is insomnia. A growing number of them look to the molecule melatonin to help them sleep, and to cope with jet-lag. Companies who make this drug report record demand.

Melatonin is a hormone produced by the pineal gland, which is about the size of a pea and lies at the centre of the brain. It regulates sleep by releasing melatonin molecules at night, and it does this in response to changes in light entering the eye. Levels of melatonin in the bloodstream reach a peak in the small hours at around 80 parts per billion (ppb), and then decline slowly, falling sharply at dawn to 10 ppb. As we reach old age, our ability to produce melatonin decreases, and night-time levels may be only half that of a young person.

Taking a 3 mg capsule of melatonin is enough to quickly raise the melatonin level in the blood, and will send you off to sleep in about five minutes. Melatonin can be bought from health food stores, where they claim it is a nutrient, although the UK Committee on the Safety of Medicines has now banned its sale as a non-prescription drug.

The European Pineal Society, while admitting that melatonin is useful in treating sleep disorders, has issued a warning. They say that there is insufficient scientific evidence for its therapeutic use in humans, and there is as yet not enough information on its possible harmful long-term side-effects. They warn that melatonin may be dangerous if incorrectly timed, and should not be taken without medical supervision. Perhaps they are being somewhat over-cautious, but there are already signs that some people are treating melatonin as a kind of wonder-drug. In the USA it is hyped as a cure-all, with claims that it can ward off cancer, heart disease, Alzheimer's, cataracts, AIDS, depression and old age. Melatonin mania has been fuelled by best-sellers such as *The Melatonin Miracle* by Walter Pierpaoli and William Regelson, who say it prevents ageing; and *Melatonin: Your Body's Natural Wonder Drug* by Russel Reiter and Jo Robinson, who claim it can counter cell damage caused by free radicals, and so may prevent cancer. There is as yet no convincing support for either theory, but that has not checked demand, and in some states in the USA melatonin now outsells aspirin. Packs of tablets sold over-the-counter in the UK are more restrained in their claims, and one brand, Bio-Melatonin, is curiously labelled 'a powerful antioxidant nutrient found in fruits', making it sound more like vitamin C than a brain chemical.

Readers wanting a more scholarly and lucidly written work on the subject should consult *Melatonin and the Mammalian Pineal Gland* by Josephine Arendt, who has researched the effects of melatonin on human biological rhythms, and developed sophisticated isotope methods of measuring the chemical in the body.

In the Department of Anatomy at Cambridge University, Drs Mike Hasting and Francis Ebling are carrying out research into how melatonin controls the body's internal clock, with some remarkable findings. They have discovered that the brain has two mechanisms related to time: one that regulates the daily, or circadian, rhythm, of our lives; and one that controls our response to seasonal changes. Both are sensitive to melatonin at the very low concentrations found naturally.

The body clock is located in the hypothalamus, while the body calendar is in the nearby pituitary gland.

Melatonin (N-acetyl 5-methoxy-tryptamine) can easily be manufactured, and when pure is pale yellow, with leaf-like crystals which melt at 117 Celsius. The pineal gland produces it from serotonin, the brain chemical that regulates mood, and this in turn is made from the essential amino acid tryptophan. They are all derivatives of indole, a simple molecule which has two rings of atoms closely joined together; one ring has six carbon atoms, and the other has four carbons and a nitrogen.

The dermatologist Aaron Lerner discovered melatonin at Yale University in 1958, and he reported that in frogs it caused dramatic changes to the colour of skin cells called melanophores—hence the name melatonin. Since then it has been found to occur in organisms ranging from single cell algae up to mammals. It has several roles: in humans it helps us to adjust our sleep patterns to the daily rotation of the planet and its annual cycle round the Sun. It also controls our body temperature, reducing it slightly during the hours of sleep. In sheep and deer, melatonin signals the breeding season, while in other animals it causes moulting, and for some it determines the time to hibernate.

There are proper uses for melatonin, for reducing jet lag in those who frequently travel across time-zones, and for increasing alertness and performance in people who have to cope with abrupt changes in shift work. Melatonin has also been used to treat children suffering from disturbed sleep patterns. Allowing over-stressed managers to get a good night's sleep may also be another legitimate use.

◆

We hope you have benefited from your visit to the exhibition in Gallery 3A, which has been included as a public service about the dangers of taking illicit and potentially dangerous substances.

GALLERY 4

HOME, SWEET HOME

An exhibition of detergents, dangers, delights and delusions

▪ KEEP IT CLEAN ▪ TRIED AND FOUND INNOCENT
▪ THE MAN IN THE WHITE SUIT ▪ ZAP THE GERMS
▪ CRYSTAL CLEAR AND CLOUDED IN MYSTERY
▪ WHAT'S IT MADE OF FROM (1)? ▪ WHAT'S IT MADE FROM (2)?
▪ DANGER IN THE HOME ▪ THE BITTER SECRET OF SAFETY
▪ ELEMENTS FROM HEAVEN (1) ▪ ELEMENTS FROM HEAVEN (2)

FEW PEOPLE today lead home-centred lives, and perhaps that's how it should be. For earlier generations the home was all important: it was a place of comfort after a hard day's work, a place to be proud of, a place of love and security for young children—and possibly a place of drudgery, boredom, quarrels and abuse. But whatever a home was, it was a place which chemistry was to transform, so that today it is cleaner, healthier, safer to live in, and with some remarkable labour-saving gadgets and entertainment facilities. It is cleaner because of detergents, healthier because of disinfectants and safer because the chemicals we use may be protected by other chemicals, as we shall see.

Throughout the 1960s, 70s and 80s the use of detergents was portrayed by some as almost wanton pollution of rivers and lakes. The chemicals that lift grease and dirt from dishes, clothes and even our own bodies were accused of causing 'eutrophication': the imbalance in rivers, lakes or inland seas that produces an excess of slimy algae or weeds. The main culprit was said to be phosphate in detergents, but even the surfactants they relied on for their cleaning ability were under a cloud, because these were produced from oil. Such was the odium under which detergents laboured that companies strove to produce 'green' alternatives that were phosphate-free. Unfortunately many consumers did not like them, but their rejection in the end hardly mattered because it turned out that phosphates were not so environmentally damaging after all.

Detergents are made up of many components, but two are worth a closer look: surfactants, which dissolve grease, and phosphates, which soften the water. It is not difficult to find something good to say about them both of them.

Portrait 1

Keep it clean—surfactants

During the Gulf War of 1991, millions of barrels of crude oil were deliberately released by the Iraqi invaders into the waters of the Persian Gulf. As usually happens, it was the local birds who bore the brunt of this ecological vandalism. Some sea mammals, such as the dugongs, were also badly affected. Black-neck grebes were the main species to be blighted, but particularly worrying was the threat to an endangered species, the socotra cormorant. The UK's Royal Society for the Prevention of Cruelty to Animals (RSPCA) was asked by the Saudi authorities to help, since their experience in rescuing oil-covered sea birds was well recognized—they had often been able to save the lives of 70% of birds given into their care. The main products the RSPCA used to remove the oil were Fairy Liquid and Co-op Green, two proprietary brands of washing-up liquid.

The secret of a washing-up liquid is its chemical surfactants. The name comes from the term 'surface-active agent'. The action of a surfactant is two-fold: it promotes surface wetting, whether of skin, clothes, dishes or feathers, and it lifts off particles of dirt and grease. Surfactant molecules are long and thin. One end, the head, is attracted to water, and the other end, the tail, is repelled by water but attracted to organic matter such as oil. The effect of the head must dominate so that the surfactant dissolves in water. When it comes up against oil it should then be able to attach itself by its tail and make that dissolve also. (The oil does not really dissolve, but becomes trapped as minute drops inside a coating of the surfactant, a state known as a micro-emulsion.)

Surfactants are in all detergents and most household cleaning aids, including soap. Soap was the first surfactant, and is still the preferred one for washing ourselves. It is made from natural oils like coconut oil or animal fats, but it has the drawback of precipitating an insoluble scum of calcium salts when it reacts with the calcium carbonate of hard water.

For washing most things, it is better to use synthetic surfactants. These do not form insoluble deposits with calcium.

The RSPCA started reclaiming stricken birds when they had to deal with the first major oil disaster in which the fully-loaded oil tanker *Torrey Canyon* ran aground off the south-west coast of Britain in 1967 and spilled all its cargo into the sea. They discovered that two brands of washing-up liquid were to be preferred, not because they removed the oil better than other liquids, but because their surfactant molecules rinsed away better. The tail ends of some surfactant molecules will attach themselves to birds' feathers and this reduces their essential waterproofing, so much so that the birds can no longer survive. Once the RSPCA team have cleaned, rinsed and dried them, birds will quickly recover and can be released back to the wild after a few days. Sadly many birds that are rescued and cleaned do not survive for long—the trauma has been too much for them.

Most detergent manufacturers rely on surfactants made from oil, and have done so for over 50 years. Shell marketed Teepol, the first of them, in the 1940s. It seems unlikely that we would ever give up the benefits of surfactants like these, and so there has been a move to finding renewable alternatives to oil-derived surfactants. Several companies have devised 'green' surfactants, which are chemically similar but are made from plant oils and other renewable resources. Companies in Japan and Germany have patented surfactants, called monoalkylphosphates, which chemists there have made from sugar and vegetable oils, and these have been marketed. Despite media acclaim, they made little impact.

However, they were not marketed in vain, and it was their possible success that made established detergent producers look again at their existing products. Not surprisingly they discovered that these in fact were rather wasteful of materials and packaging. Soon they were offering concentrated versions, in smaller packs, which also washed just as well but at much lower temperatures such as 40 Celsius, thereby saving the customer's and the world's energy resources. What puzzled those in the industry was that many people chose to use the same volume of powder as previously, so getting fewer washes per pack of detergent. The sales of concentrated detergents eventually declined as the customers returned to the older type of 'big box' powders, although these had by now become much more sophisticated and could also wash at lower temperatures. (They actually wash even better at higher temperatures.)

Whether they are made from oil or from renewable resources, these surfactants are still very similar molecules. Natural surfactants, in other words those made by living things, are much more complicated. Some are essential to life. It is a curious irony that biologists were to discover that humans produce surfactants when they studied the effects of war gases on the lungs in World War II. Many premature babies used to die because their lungs lacked surfactant, which is needed to keep the small air spaces open. The surfactant molecules do this by overcoming the surface tension of water, which is the internal force within water which has to be reduced in the lungs to allow the air spaces to open. In some premature babies the natural surfactant is lacking, their lungs collapse, and they die.

Colin Morley, of the Addenbrookes Hospital, Cambridge, has researched this human surfactant, which is made up of lipids and proteins. According to Morley, natural surfactants have a remarkable chemical ability. When we breathe in, our surfactants reduce the surface tension of water, allowing our lungs to expand easily. However, it is when we breathe out that their uniqueness is revealed. The surfactants *solidify*, and this prevents the very fine air passages in our lungs from collapsing. Morley has been making a modified version of the natural variety and using this in place of the natural surfactant to keep babies breathing until they can make their own. This has halved the death rate of premature babies.

The head of most surfactant molecules attracts water by carrying a negatively charged group of atoms. In soap this is a carboxylate group; in synthetic surfactants it is a sulphonate group; in human surfactants it is a phosphate. Some surfactants can be positively charged, and some carry no charge at all and rely on oxygen atoms to attract water. The negatively charged surfactants account for over half the sales of all surfactants, and are the cleaning component of the washing-up liquids used to rescue birds.

The tails of surfactants are long chains of hydrocarbons, which is why they are so good at penetrating oils and grease—'like attracts like'—leaving the head of the surfactant still dissolved in the water and able to pull the oil and grease into the water also. However the hydrocarbon tails of surfactants once had serious environmental implications. The early synthetic surfactants were made from propylene, which, as we shall see in Gallery 5, is a polymer of carbon chains with methyl side-groups. Unfortunately these methyls prevented bacteria in sewerage plants from digesting the surfactant molecules, and so they passed out with the waste water causing the foaming rivers of the 1960s. Chemists then re-designed the polymer chains, and by making them from ethylene, which has no side-groups, they produced surfactants that bacteria can digest.

Oil slicks at sea can safely be left to the natural dispersion of winds, waves and storms, and this does far less damage than attempts to disperse them by spraying with surfactants. The negatively charged surfactants which were once used for this caused more damage to marine life than did the oil itself.

Portrait 2

Tried and found innocent—phosphates

Phosphates have not been very popular with some campaigners who blame them for polluting rivers, lakes and inland seas. (Phosphates pose less of a threat to the oceans where they are quickly taken up in the food chain and eventually fall to the deep.) Some of this polluting phosphate came from artificial fertilizers and from domestic sewage, but most came from the detergents used to wash clothes and dishes.

Sodium tripolyphosphate (STPP), once the main ingredient in detergents, softens water by sequestering calcium, and it keeps dirt in suspension once it has been washed off clothes. In some washing powders, up to a third of their weight was phosphate. In the 1980s detergent manufacturers reduced the amount of STPP in their products, and some went all the way so they could label their products 'phosphate-free'. In many parts of Europe phosphate-based laundry detergents disappeared from supermarket shelves as a result of environmentalist action, although dishwasher powders still continued to rely on STPP.

The trouble starts when STPP enters the sewers, and adds to the phosphate from human excreta and industrial waste. Some of this phosphate was removed by conventional sewage treatments, but most ended up in rivers. Many cities invested heavily in improved water treatment plants to remove the offending phosphate. In theory this may not be a bad thing to have done, although it was expensive in economic terms. New sewerage technology allows phosphate to be recovered and recycled, and we may yet find detergents once again proudly boasting that they contain (recycled) phosphates. One day we may even be washing our dishes with the help of phosphate reclaimed from sewage. We might also be using it to keep our chicken salad germ-free (see below); we could even be recycling it back into colas (see Gallery 1).

In the past, rivers like the Rhine and the Great Lakes of North America were the ones to suffer eutrophication, and yet detergent phosphate may not be as environmentally damaging as once believed. This was the conclusion of *The Phosphate Report*, written by Bryn Jones, a former director of Greenpeace, and Dr Bob Wilson. They carried out a full environmental audit of STPP, and compared it with the alternative detergent component, zeolite, which is an aluminium silicate and regarded as environmentally benign. The report is a cradle-to-grave analysis of these chemicals, and takes into account all environmental costs including mining the raw materials, industrial production, energy input, transportation costs, use by consumers and environmental pollution. The net result is that there is little to choose between the despised phosphates and the praised zeolites, and improvements in sewage treatment might well tilt the balance in favour of a return to phosphate in the next century.

In Sweden phosphates are actively encouraged, while in The Netherlands they already reclaim waste phosphate as calcium phosphate, and one chemical company has shown that this can be recycled into STPP. Reclaiming phosphate from sewage has other benefits, not least because it also extracts heavy metal impurities, such as cadmium and chromium, leaving behind a sludge that can be used as fertilizer for farmland.

Calcium phosphate is the form in which most phosphate occurs in the Earth's crust, and it is found in vast deposits. It has been used for making fertilizers for over 150 years. Originally bone meal was used, which is calcium phosphate, but this provides relatively little fertilizer today and is only used by gardeners and 'organic' farmers. Most crops are grown using calcium superphosphate, a more soluble form made by treating calcium phosphate rock with sulfuric acid.

Some detergent phosphates have re-emerged under new guises. Trisodium phosphate was for many years the miracle ingredient in Flash, the bestselling domestic surface cleaner. Now this phosphate is being used to remove food-poisoning germs from raw chicken. Surface bacteria, such as salmonella, are washed away when a carcass is sprayed with a solution of trisodium phosphate, which works by removing the layer of surface fat that covers the germs and allows them to cling to the meat. The process got approval in 1992 from the FDA in the USA. Many of the thousands of cases of salmonella food poisoning which occur every day can be traced to infected poultry meat, and many are being prevented with a little help from phosphates.

Portrait 3

The man in the white suit—perfluoropolyethers

In 1951 Alec Guinness starred in the film *The Man in the White Suit*, which was the story of a young chemist who discovered a fabric which never wore out and never needed washing. The plot revolved round an unholy alliance of textile mill bosses and union leaders who were determined to see that his threat to their livelihood was sabotaged. In the end the new fabric defeated itself by falling apart anyway. What was comedy and satire then could become reality in the next century. There already exist chemicals that can be used to treat surfaces so that they repel dirt, but whether they will ever be used to treat fabrics is unlikely—but not impossible.

The chemicals are the perfluoropolyethers (PFPE for short), and they have already been used to protect buildings and are added to some polishes. PFPEs are not new, but until the 1980s they were rather expensive (they can cost up to £300 ($500) per litre). Their use could only be justified for vehicles in space, where conditions are so extreme that conventional oils cannot cope. The PFPEs have a unique set of properties that make them ideal as lubricants: they spread evenly, they are not affected by high or low temperatures, they can cope with corrosive chemicals such as acids and oxidizing agents, and they are non-flammable.

These desirable features derive from the molecular structure of PFPEs, which consists of chains of carbon atoms to each of which is attached two fluorine atoms. These chains are linked together through oxygen atoms to form even longer chains. The fluorine atoms provide a hard coating which protects the chain while the oxygen atoms give them flexibility. The result is a polymer that is impervious to any attack, and which does not mix with anything else but its own kind. Unlike common oils, PFPEs do not penetrate plastic surfaces so they make superb lubricants for videotapes, rubber gloves and condoms.

But lubrication is not the only advantage of PFPEs. Their exclusiveness makes them good barriers to dirt and grime. Moreover, they are perfectly safe biologically and environmentally. They score over other compounds such as polyurethanes and silicones when used as protective coatings, because they do not discolour in strong sunlight and they discourage bacterial and fungal growth.

PFPE was discovered 20 years ago by the Italian chemists Dario Sianesi, Adolfo Pasetti and Constante Corti. They are made by reacting tetrafluoroethylene or hexafluoropropylene with pure oxygen gas and ultraviolet light at −40 Celsius. This forms peroxides which when reacted with fluorine gas at 200 Celsius forms the PFPEs with a range of chain lengths which can be separated into lighter and heavier oils by distillation. The lighter ones are used in testing electronic equipment because they can absorb heat while being non-conducting. A television set or PC can be immersed in them and still continue working. The heavier oils with longer chains are ideal for protecting buildings. The short chains have also been test-marketed in products such as shampoos, bath oils, soaps and suntan creams because they are odourless, colourless, transparent, non-toxic, non-irritant and give a satin-like feel to skin and hair.

PFPE has been used in Italy to protect historic buildings. Corrosion is now a serious threat faced by many ancient monuments, cathedrals, palaces and other great buildings in traffic-congested cities, but tests over many years have demonstrated that PFPE, sprayed on to stone and marble after it has been cleaned, will preserve the surface against all further corrosion. The high fluidity of PFPE means that it penetrates even the smallest crevices. Whole cathedrals, such as those at Syracuse in Sicily and Lucia in Tuscany, have been renovated and protected with PFPE. Whether it can do the same for domestic surroundings remains to be seen.

Portrait 4

Zap the germs—sodium hypochlorite

Environmental pollutants may seem to be an insidious threat to our well-being, but they are as nothing compared to the natural threats to our health from disease-causing viruses and bacteria. Destroying these is part of the battle we wage to keep healthy, and what better ammunition is there than the humble bottle of bleach for protecting ourselves, our families and our homes? Bleach, as advertisers claim, really does kill all known germs, and has been doing so for almost a century. Nor is it possible for any germs to evolve to withstand its attack, because it damages them wherever it touches them. Viruses and bacteria may harm us once they have gained access to our bodies, but our first line of defence is to make sure that they are not even around, and that is where bleach can help.

Bleach is made from chlorine gas. It was first used to disinfect tap water at Maidstone, England, in 1897 during an outbreak of typhoid, and its efficacy was confirmed when it helped to control another epidemic at Lincoln a few years later. Eventually chlorine became the method of purifying drinking water throughout the British Isles, and today most of the developed world uses it. Nevertheless, it is frowned upon by some environmentalists because it reacts chemically with other matter that is present in water to form traces of compounds suspected of being carcinogens. Bleach is also outlawed by some local authorities who forbid its use in schools and even in hospitals because of the fear of releasing dangerous chlorine fumes. This can sometimes happen if cleaning staff are careless.

Bleach is made by bubbling chlorine up a column, down which trickles a solution of the alkali sodium hydroxide. The two react to form sodium hypochlorite, a molecule which has an oxygen atom and a chlorine atom joined together. Hypochlorite is a strong oxidizing agent, and is stable for several months provided it is not exposed to heat, sunlight or metals. Because it is made from chlorine, ordinary bleach is sometimes wrongly called chlorine bleach, to distinguish it from peroxide bleach which is a solution of hydrogen peroxide. In chlorine bleach there is no free chlorine gas as such, unless the solution becomes acidic, as we shall see. The bubbles which rise from such bleach when it is being used are normally bubbles of oxygen gas.

Throughout the world millions of tons of chlorine are manufactured each year, and much of it is used to disinfect water or is turned into hypochlorite bleach for use in the home. Water can be chlorinated either with bleach or with chlorine gas itself; in both cases the result is a dilute but highly effective solution of hypochlorite. Bleach is also used industrially to remove ink from recycled paper and to bleach cotton, one of the first uses to which it was put.

Viruses and bacteria are extremely sensitive to oxidation and an attack by even weak hypochlorite solution normally kills them. Hypochlorite will keep water free of germs at very low concentrations and for a long time, which is why it is still preferred over short-lived oxidizing agents such as hydrogen peroxide and ozone. Bleach is ideal for sterilizing kitchen surfaces, soiled garments, sinks and toilets. Thickened bleach is produced by adding a surfactant which also helps in the cleaning action.

There is still no real substitute for hypochlorite bleach as a disinfectant, so why is it being actively discouraged? One reason is its ability to convert organic residues in water into organochlorine compounds which, some argue, are a long-term threat to public health. There is a little epidemiological data which seems to support this: for example, a report in 1992 claimed there were a few extra cases of bladder and rectal cancer per million people in areas where the drinking water was heavily chlorinated and taken from rivers, compared to areas where lightly chlorinated drinking water was extracted from springs and wells. However, the difference was not compelling, and fell far short of proving that anyone drinking chlorinated water is putting their health at risk.

US river water has on average a mere 50 ppb of organochlorine compounds; in the UK it is much less than this. Both the US Environmental Protection Agency and the UK authorities set a limit of 100 ppb for these types of chemicals in drinking water. Some organochlorine compounds have caused cancer among those heavily exposed to them in industry, but at the low levels at which they are present in chlorinated water the risk of them affecting anyone is negligible. The most common organochlorine residue in water is chloroform, but even at 100 ppb you would consume only 3 g (a tenth of an ounce) of chloroform in a lifetime from this source. This amount should be compared to the chloroform that was actually taken as a medicine in the first half of this century. Chlorodyne, a patent cure-all, contained 14% chloroform, and a single dose of this would provide as much chloroform as you would get from drinking water for 75 years.

In 1991 the International Agency for Research on Cancer (IARC), which is part of the World Health Organization, published an evaluation of the risks posed by organochlorines in drinking water. It concluded that there was insufficient evidence to cause alarm. The IARC report said that if there are any health risks they are very low, and they have to be offset against much greater health risks associated with drinking unchlorinated water. But this cautionary note came too late. Others were taking the threat seriously and, alarmed by the supposed dangers of organochlorines, the Peruvian government heeded the EPA and stopped chlorinating drinking water in 1991. As a result there was an outbreak of cholera with over a million cases, and 10 000 died.

However, organochlorines are not the reason why bleach is discouraged in schools, hospitals and workplaces. The reason behind this ruling is that it may release chlorine gas if it is used wrongly. Several people are rushed to hospital each year because of such accidents, and bleach is often cited by safety officers as a common chemical that can be dangerous. Generally it is cleaning staff who are most likely to be affected if they try to save time by using a bleach and a descaler together. Descalers are strong acids and work by dissolving the calcium carbonate that builds up on surfaces, sinks and toilets in hard-water areas. As well as neutralizing the scale, descalers can neutralize the alkali in bleach and make it acid enough to convert hypochlorite back to hazardous chlorine gas. Some descalers even contain hydrochloric acid, which is doubly dangerous because it releases some of its own chloride as chlorine gas. Here the real culprit is not bleach, which has saved millions of lives in its time, but ignorance. If people know a little chemistry then it is safe to use bleach, which really can kill germs and make kitchens and toilets safer places.

Parents worry about bleach because they imagine it is dangerous if it touches the skin, and children may play with it and may drink some. The smell alone should deter children from drinking it, but they may try to. The threat is not as dangerous as you may fear, and if a child is suspected of having drunk some bleach, then the immediate first aid should be to get them to drink lots of water, preferably with bicarbonate of soda dissolved in it to neutralize stomach acid. Other household chemicals also pose a threat, but parents can protect their children against these by looking for those which have a bittering agent added to them. Curiously, the one household chemical that does not take a bittering agent is bleach. (Later in this Gallery we will be able to admire the molecule which is the bitterest substance on Earth.)

Portrait 5

Crystal clear and clouded in mystery—glass

In the modern home we rely heavily on glass for drinking vessels, dishes and containers. One of the many benefits of glass is that it cleans easily, and can be sterilized by boiling water, and generally it is strong enough for us to knock it and wash it without worrying too much about it breaking. In previous ages glass was fragile and dangerous, and was saved only for use on special occasions. But things might have been very different had not the inventor of a type of toughened glass met a sad fate. We don't know his name, but he lived about the time of Jesus Christ. He boasted that his glass was unbreakable, and he was rightly proud of his invention. News of his remarkable achievement reached the Imperial Court of Tiberius, who was Emperor of Rome from 14 to 37AD.

Tiberius' reign was marred by political scandals surrounding his chief henchman, the praetorian prefect Sejanus, and the Emperor was notorious for his sexual perversions as a paedophile, in which he indulged at his palace on Capri. Despite his own unpopularity, Tiberius ruled at a time of great prosperity—at least for the citizens of Rome, if not for their subjects and slaves. We can see this reflected in the glassware of the period. Indeed, some of the vessels that have survived are still among the most beautiful works of art ever wrought in glass, such as the layered blue and white Portland Vase in the British Museum. This particular vessel was deliberately smashed by a youth in 1845, but it has twice been painstakingly reassembled, and the effort was worth it, such is the quality of the workmanship and artistry that it represents.

Roman glass, like most glass throughout history, was easily shattered.

The only exception, before the advent of Pyrex glass, was seen by the Imperial Court when our intrepid craftsman brought one of his creations, a beautiful transparent vase, to show to the Emperor. He then deliberately dropped it on the floor, and to everyone's amazement it did not break. The onlookers were amazed, but some were alarmed, and others suspected he must be using sorcery. The Emperor kept his nerve and asked the glassmaker about his wonderful new glass. What was it made from? Who else knew the secret of how to make it? The glassmaker said that it was made from martiolum, and boasted that he alone knew the recipe. On hearing this, the Emperor ordered his immediate execution, and the glassmaker's secret died with him. Just to be on the safe side, Tiberius ordered the destruction of the man's workshop as well. The reason he gave was the understandable one of protecting the value of the palace's existing investments in glass artefacts and tableware—not very different from the motive of those who opposed the man in the white suit in the 1951 film.

The ancient writers Pliny and Petronius, who reported the incident, called the glass *vitrium flexible* (flexible glass) and said that it was made from martiolum. This material we cannot now identify. Or can we? I believe it must have been a form of sodium borate, which is a complex compound of sodium, boron and oxygen. This is the key ingredient of borosilicate or Pyrex glass, and is the reason why this can withstand hard knocks, sudden temperature changes and chemical attack. The borate component of the glass is essential because it can adjust its chemical bonding to absorb these sudden energy changes.

Pyrex was discovered (or rather 'rediscovered', if my theory is correct) in the 1880s by Otto Schott, Karl Zeiss and Ernst Abbé in Germany. It was marketed as Pyrex ovenware by the Corning Glass Company in 1912, and soon became a part of every home—and it is still the most important material in chemistry labs. Ordinary, untoughened glass is made from sand (silicon dioxide), limestone (calcium carbonate) and soda ash (sodium carbonate) and has a typical composition of silicon oxide 70%, sodium oxide 15%, calcium oxide 10%, with other oxides 5%. But add to such a glass a little sodium borate and you transform its properties completely. To make borosilicate glass you typically need to add around 10% boron oxide.

Could our unknown glassmaker have stumbled upon a type of Pyrex glass? If the reports are correct, and he really had made a shatterproof vase, then he must have added some borate, of that we can be sure, because there is no element other than boron that can give glass this resilience to knocks and thermal shocks. He probably used borax, the most common natural mineral of sodium borate, but where did he get it from?

The only source of borax in the ancient world was in faraway Tibet, where it crystallizes from Lake Yamdok Cho, south of Lhasa. From there it was exported to the Near East and Europe right up to the end of the eighteenth century. It was used as a flux by goldsmiths, but a little borax went a long way so that importing it from such a distance was an expense the trade could bear. There is no evidence that Greek and Roman goldsmiths used borax for this purpose, and even though there are references to borax in ancient Babylonian, Egyptian and Roman literature we cannot know if this really was sodium borate or another salt. Perhaps some of it was borax, and it may be that our intrepid glass chemist got hold of a little. He may even have stumbled upon his own supply.

Unknown to the Romans at the time, there were large deposits of borax within the borders of their Empire, for example in Asia Minor (Turkey), which is today a major exporter of the mineral. Much nearer Rome was another source, at Maremma in Tuscany. This deposit was mined in the nineteenth century, and for 30 years, up to 1850, Italy was the world's largest producer of borax. Perhaps our glassmaker experimented with the boric acid which he might have collected from around the steam vents of the Maremma region, and which would behave in much the same way as borax if added to molten glass. The name he gave his material was martiolum, and this may have been named after the locale where it was found, which is a further clue that he might have used a boron-containing compound from this region.

Sadly, our early glass chemist took his secret with him to his grave, and the world had to wait nearly 2000 years before anyone again experimented and added borax to molten glass. Had Tiberius rewarded the glassmaker with a research budget, in the expectation of further discoveries, he might well have founded the Imperial Pyrex Company and an industry that was not only wealth-creating, but that also improved the health of his subjects. Lead pewter tableware and goblets were what the upper-class Romans used, and these are reputed to have contributed to the downfall of Rome by increasing the amount of lead in their diet to levels that could have affected fertility. We shall look at this theory again in Gallery 8.

Of course, even if the Romans had developed a toughened glass it would still not have changed their domestic surroundings in quite the same way as new materials have been able to transform ours in this century.

Portrait 6

What's it made from? (1)—ethyl acrylate

On Friday 3 May 1991 the cargo vessel *Nordic Pride* was caught in a storm in the North Sea. Two tanker trailers each holding 24 000 litres (5000 gallons) of ethyl acrylate were washed overboard. They came ashore on the east coast of England, near Kelling in the county of Norfolk, on 6 May and were discovered by a man walking his dog. One of the containers was leaking slightly through a valve. That same day local residents reported a strong garlic-like smell, which is not surprising since ethyl acrylate has a vile odour.

The Norfolk emergency services swung into action when they discovered they were dealing with ethyl acrylate because it is classed as a potential health hazard. (It can be dealt with by diluting it with large amounts of water, which is the best method of decontamination.) They mounted a full emergency operation, and set up a two-mile exclusion zone inside which the police advised people to leave their homes. Others within a ten mile radius were told to stay indoors and close their windows.

The ever-vigilant media picked up the scent of what appeared to be a major chemical disaster and reported toxic fumes affecting a wide area. The local hospital dealt with 48 people who thought they had been affected by ethyl acrylate vapour, but no one was hospitalized. Meanwhile a specialist disposal company siphoned off the ethyl acrylate and took it away. Only a little was missing. Media interest waned immediately—there were no clouds of toxic fumes engulfing innocent people—but the lack of follow-up must have left some people with unanswered questions. What was this strange chemical that had washed up on the beach? Why was there such a large amount of it? What was it used for? Was it really dangerous?

Ethyl acrylate is a raw material for the chemicals industry, and is used to make acrylic polymers. It is a pungent, colourless liquid which boils at exactly the same temperature as water, and it is very irritating to eyes and lungs. Unpleasant as ethyl acrylate is, the raw material bears little relation to the finished product, which is a harmless polymer we handle many times a day because it coats walls, floors, steel, paper and leather. It is also used as the binder for disposable non-woven fabrics, such as nappy (diaper) linings and headrest covers in aircraft. Over three million tons of acrylic polymers are manufactured worldwide each year, and production is growing at more than five per cent per annum.

Most ethyl acrylate ends up in paint. A little is used in domestic water-based emulsion paint, where its higher price is worth paying for its resistance to damp in kitchens and bathrooms. Most is used in industry as solvent-based paint for coating household appliances and car bodies, which are sprayed with the paint which is then baked fast. The polymer is ideal for covering surfaces because it is flexible, tough, stands up well to regular cleaning, and has excellent resistance to bad weather and strong sunlight. Ethyl acrylate is perfectly safe when it has been polymerized and as such is now being used to coat the inside of aluminium cans so that their contents, especially fruit acids in drinks, do not react with the metal and contaminate the product.

Acrylate polymers do their job by binding the pigments in paints and ensuring they stick to the surface being decorated. Ethyl acrylate is added to a formulation if the paint is destined for surfaces where it must have a bit of flexibility, such as on textured wallpaper. In such paints the ethyl acrylate is co-polymerized with the more rigid methyl methacrylate which is the main component. Generally acrylic paints are used to cover metal surfaces, such as fridges, washing machines, dishwashers and cars, and they give them a lacquer-like finish. Indeed, they are so strong that they can be applied to steel sheet before the metal is stamped into component parts. Good as they are, acrylic paints have been criticized because of the solvents needed to apply them, but new environmentally friendly acrylate paints have been developed which are water-based and are already being used by European and the US motor manufacturers to spray car bodies.

Acrylic is the general name given to a whole class of polymers, not just those made with ethyl acrylate. Methyl methacrylate plastic is better known as Perspex and Plexiglas. Another one, methyl cyanoacrylate, is the basis of super-glues. This is sold as the unpolymerized compound, and has been designed by chemists only to polymerize when it is exposed to the air. Its superior stickability derives from the long chains of polymers that form across the surfaces it is joining, binding them together as if they were part of the same piece.

Portrait 7

What's it made from? (2)—maleic anhydride

As with ethyl acrylate, you will probably never come into physical contact with maleic anhydride itself. You may never even have heard of it before entering Gallery 4, but from the moment you get out of bed in the morning you can hardly avoid handling materials made from it. You may have showered surrounded by them, sprayed them on your hair, and drunk them for breakfast. In your car you rely on them to keep the engine running smoothly, and to protect you if you bump the car in front. They are there as you sweeten your coffee, eat a biscuit and read this book. And you may even say good night to them as you remove your false teeth.

The Huntsman group of companies, based in Salt Lake City, Utah, USA, is the world's largest producer of maleic anhydride, and is a leading plastics and resins manufacturer with annual sales of around $2 billion. In 1993 Huntsman bought the maleic anhydride division of the giant US chemical company Monsanto, and Huntsman's senior vice president claimed that they had acquired a remarkable facility which made maleic anhydride from a gas that is 98% air. Nor was this an idle boast: in the chemical plant they had bought, the maleic anhydride was made in a matter of seconds by passing air containing 2% butane, an alkane hydrocarbon derived from oil, over a mixture of metal oxides such as those of vanadium or molybdenum, at 350 Celsius. More than 200 000 tons of maleic anhydride are manufactured each year, and it is one of the few chemicals whose production increases year by year at around 10%.

Maleic anhydride can be made from a variety of chemicals, such as benzene or butane, simply by passing a stream of air containing these vapours over a metal oxide catalyst. Maleic anhydride is a simple molecule with a ring of five atoms: four of carbon and one of oxygen. The carbons either side of this oxygen each have another oxygen attached to them as well. Pure maleic anhydride is a white crystalline solid which melts at 53 Celsius. It is toxic and irritating to the skin, and it is never used as such; but it is converted to other chemicals that are turned into a wide range of products: plastics for bathrooms and car bumpers; additives for vinyl flooring and mouthwashes; and food ingredients such as artificial sweeteners and baking powder. Other outlets for maleic anhydride are engine oils, where polymers made from it keep oil viscous when it gets hot, and inks, where it acts as a binder.

It may seem strange that the same chemical can end up in such a range of products, but this is because maleic anhydride is only the starting material. It can undergo many chemical reactions, each one changing it to a more sophisticated material with a new range of properties. Nor should we wonder at this kind of chemical transformation. A starting material with which we are more familiar is sugar, and this too can undergo remarkable changes. We can turn it into a soft brown caramel, a thick syrup, or a clear hard boiled sweet; we can spin it into fibres as in candy-floss; or cast in into a sheet of glass, to be jumped through dramatically on a film set.

The largest end-use for maleic anhydride is polyester resins, and more than half of maleic anhydride production ends up this way, as hulls for boats, shower cubicles, and marble-style work surfaces. It is ideal for these because it is lightweight, strong, safe and does not corrode. The resin is made chemically by reacting maleic anhydride with a double alcohol such as propylene glycol to form a polyester. Unlike the polyester used for textiles, this kind has its polymer strands cross-linked, making it rigid. The resin can be further strengthened with fibreglass, and if it is sprayed into a mould then we have a quick and cheap way of building not only boats, but also film sets, theatre props and even theme parks. The most striking use of the polyester is in making the water-clear resin for paperweights in which souvenirs, coins and other items are embedded like flies in amber.

Anhydride is a chemical term that refers to an acid from which the elements of water, i.e. two hydrogen atoms and an oxygen atom (H_2O), have been abstracted; the word is derived from the Greek for 'without water.' When this is done to maleic acid the result is maleic anhydride. The process can be reversed, and adding water to maleic anhydride reconstitutes maleic acid. Again, this is not sold as such, but is converted to related chemicals that occur naturally: malic acid, tartaric acid and fumaric acid.

Fumaric acid has the same chemical composition as maleic acid, but with a different twist to the molecule's structure. It is essential to plant and animal tissue respiration. Fumaric acid is used as a flavouring agent and as an antioxidant in some foods, mainly instant desserts and cheesecake mixes. It is perfectly safe, and although it occurs naturally in many plants it is not present in quantities that make extraction an economic process. Instead it is made industrially, either from glucose by the action of fungi, or from maleic acid.

Malic acid is made by heating maleic anhydride with steam under pressure, and is used to make marmalades, jellies and fruit drinks. Human blood naturally has 5 ppm of malic acid. The sour taste of unripe apples is due to malic acid, and these contain 1%, most of which disappears as the fruit ripens (although some varieties, and particularly those known as cooking applies, have high levels even when ripe). Rhubarb and gooseberries have as much as 2% malic acid. Adding malic acid to fruit drinks gives them a refreshing tartness.

The same effect, but with a 'rougher' taste, can also be achieved with another natural acid, tartaric acid. This too is now mainly made from maleic anhydride, by reacting it with hydrogen peroxide. Some tartaric acid is extracted from wine sediments, which was where most of it used to come from. Tartaric acid is used in baking powders as its potassium salt, and this has been sold for centuries under the name 'cream of tartar'. Maleic anhydride is not only involved in making foods taste tart; it is also responsible for making them sweeter, because it is one of the starting chemicals for the artificial sweetener aspartame (better known as NutraSweet).

In the USA a derivative of maleic anhydride is used as an agricultural chemical. Called maleic hydrazide, it encourages fruit to ripen more quickly, but it also slows down leaf growth: for example, it is used to make grass grow more slowly, especially on highway verges. It is also used to prevent tobacco sprouting new leaves and thereby reducing the quality of existing leaves. In the UK it can be used on potatoes and onions to stop them sprouting during storage. The amounts needed are tiny, and there is no danger to health from eating fruit and vegetables that have been treated this way.

Portrait 8

Danger in the home—carbon monoxide

The two portraits we have just looked at, ethyl acrylate and maleic anhydride, are to be admired because they make our homes a safer environment. Like most chemicals, they are low on the list of domestic risks, and the real dangers that we face around the home are from accidents such as falls, electric shocks and cuts from sharp implements. But not all chemicals are as innocent as those made from ethyl acrylate and maleic acid, and one in particular can be deadly. It is not one that we buy, but one that we make accidentally.

On the first Sunday in spring 1993, a young couple, Michael and Deborah Mason, set off from the London suburb of Fulham to spend a weekend at their cottage in Muddiford, North Devon. With them were their two young sons, Christopher, aged four, and Jeremy, aged two. Meanwhile in Muddiford, spring was also in the air and a pair of birds had built a nest and were waiting for a clutch of eggs to hatch. Unbeknown to the Masons, the birds had chosen to build their nest in the flue of the cottage's gas heater, and they had effectively blocked it. That weekend Michael, Deborah, Christopher and Jeremy died of carbon monoxide poisoning.

Carbon monoxide targets the haemoglobin in red blood cells and renders it useless for carrying out its essential function of transporting oxygen around the body. Without oxygen we quickly die, and our brain dies first. Carbon monoxide (chemical formula, CO) is a colourless, odourless, highly toxic gas. We are all exposed to it because it is present in trace amount in the atmosphere, with the highest levels in cities, where it comes mainly from car exhausts. Most of the carbon in a fuel ends up as carbon dioxide (CO_2) with two oxygens per molecule, but a little of the fuel in an engine or boiler may find itself without enough oxygen for complete combustion, and the carbon then ends up as carbon monoxide with only one oxygen per molecule.

The carbon monoxide in the air we breathe can tie up 5% of the haemoglobin in our red blood cells, and if we smoke as well this figure can be as high as 10%. Tiny amounts of CO in the blood are perfectly natural because metabolic degradation of haemoglobin produces CO and each day we generate about 10 mg of CO. This is enough to convert 0.5% of the haemoglobin to the CO form, and it has no effect on the blood's ability to transport oxygen around the body. Indeed, tests on mice show that a little CO seems to spur the unaffected haemoglobin to pick up more oxygen than usual, and may actually aid respiration. However, if it reaches 30% we experience the symptoms of carbon monoxide poisoning: drowsiness, headache, dizziness and chest pains. Only 1% of CO in the air converts over 50% of the haemoglobin in the blood to the useless CO form and will cause death within an hour. Nor is the potential victim aware of any difficulty in breathing, but when the 50% level is reached then oxygen transport by the blood suddenly ceases. (Some species are immune to its toxic effects and cockroaches can survive in an atmosphere of 80% CO and 20% oxygen.)

Victims of carbon monoxide poisoning turn bright pink from the carboxy-haemoglobin in their blood. If they are caught in time they can be saved, and treatment is simple: fresh air, or better still, oxygen. According to John Timbrell's *Introduction to Toxicology*, it takes about four hours to halve the level of carbon monoxide in the blood, but only one hour if the victim breathes pure oxygen. Even so a person who survives carbon monoxide poisoning may well suffer permanent damage to their heart and brain.

The worst accidental mass poisoning by carbon monoxide occurred on 2 March 1944 at Balvano, Italy, when a packed train stalled in the Armi Tunnel and 521 people died. The build-up of carbon monoxide can be very rapid if coal (or oil or gas) is burnt in an inadequate supply of air, and this is what happened in the tunnel. The deadly gas first overcame the engine driver, and then slowly filled the confined space until everyone on the packed train succumbed.

Generally people are most at risk of carbon monoxide poisoning when they move into new accommodation or holiday apartments, when they may not realize that previous occupants have sealed ventilators or that a flue or chimney has become blocked. The danger signs are gas that burns with a yellow flame, and sooty marks or fumes which show that a heater is starved of air. Every year several holidaymakers die from carbon monoxide poisoning, often in a bathroom with the door and window closed, where the hot water is supplied by an unventilated gas boiler in the room itself.

The largest human source of carbon monoxide is car exhausts, and this can result in levels as high as 50 ppm (0.005%) in very heavy traffic. Exposure to 120 ppm for an hour is considered the upper limit of safe exposure in the USA, and even 75 ppm can result in a 30% saturation of the blood and detectable body changes. Nor does closing car windows protect us from the carbon monoxide emitted by other vehicles—the level of carbon monoxide inside a car can be more than double that on the outside.

Gases from a car's exhaust contain 4% carbon monoxide when it is being driven, but this rises to 8% when the car is idling. Suicides who have attached a hose to the exhaust and have piped the fumes into the car die very quickly, just like the people who put their head in the gas oven when homes were supplied with coal gas, which contained 8% carbon monoxide.

Carbon monoxide makes up between 0.5 and 0.2 ppm of the atmosphere, and there are about 500 million tons of the gas circling the globe. The residence time of carbon monoxide in the atmosphere is about two months. During the 1990s the total amount has been declining, yet each year human activity adds 450 million tons of carbon monoxide to the atmosphere. Roughly half the carbon monoxide humans produce comes from the burning of fossil fuels, and half from the burning of wood and straw and from the clearing of forests. There is also a lot of carbon monoxide generated naturally from the oxidation of organic molecules in the atmosphere, such as methane and volatile hydrocarbons. Where all the carbon monoxide ends up is still not completely resolved, but soil is a natural 'sink' for this gas, and soil micro-organisms are thought to absorb the largest amounts.

The International Geosphere Biosphere Programme is studying all gases in the atmosphere which come from natural sources. Carbon monoxide is being measured at sites around the world, but at concentrations of less than 0.1 ppm it is not easy to monitor. Analysts are trying to find better ways of measuring carbon monoxide at these low levels using such techniques as tunable lasers, photoacoustic spectroscopy and infrared spectroscopy.

Dangerous though it is, carbon monoxide is manufactured on a large scale by the chemicals industry, and most of it is made by reacting methane gas with steam. The mixture of hydrogen gas and carbon monoxide which this produces is known as syngas, short for synthesis gas, most of which is used directly to make methanol, a liquid with many applications. Some is blended with petrol to give a cleaner burning motor fuel with the pleasing advantage that it leads to less carbon monoxide being emitted from the car's exhaust. A lot of the methanol is reacted with more carbon monoxide and converted to acetic acid and acetic anhydride, chemicals which are then used to make plastics, paints, printing inks, painkillers and pickled onions. There is a portrait of methanol in Gallery 7.

Protecting ourselves against carbon monoxide is not easy, because we do not know that we are being affected by it. It can be sensed by means of a carbon monoxide detector, and if you have a room in which there is a gas fire, then you should fit such a detector in the same way as you would fit a smoke detector.

Portrait 9

The bitter secret of safety—Bitrex

Despite the watchful eyes of parents and grandparents, every year thousands of children need medical attention because they have drunk some household chemical. Many need hospitalization, but luckily very few die. There are two ways to reduce the risk to children from this danger: the first is to ensure that all containers holding dangerous chemicals have child-resistant closures; and the second is to put something into the household chemical which will make it taste so unpleasant that any child taking a mouthful will immediately spit it out.

Accidental poisoning occurs in the bathroom, kitchen, garage or garden shed when young children find a bottle of liquid, often attractively coloured, and decide to drink some. Commonly consumed liquids include shampoo, bleach, hair conditioner, paint thinner, paint stripper, lavatory cleaner, paraffin, insecticides, disinfectants, rat poison and solvent alcohol (methylated spirits). Hospital records show that the most dangerous substances for a child to ingest are turpentine, paraffin, caustic soda, alcohol (as in scent and after-shave) and paint stripper. The extent to which a parent will panic on discovering what their child has done depends upon his or her knowledge of what the chemical is. They might believe that bleach is highly dangerous while methylated spirit is not, whereas the opposite is the case. They may be more alarmed by disinfectant than by shampoo, but in any case they should seek medical help, and the quicker the better. Happily most such emergencies are false alarms, and most children who are rushed to medical centres or hospitals need reassurance rather than treatment. Even treatment may consist of not much more than giving the child soothing stomach medicines to ease an inflamed gut, and lots of water to drink, to help the body excrete the chemicals. Only rarely is it necessary to use the popularly feared treatment of stomach pumping. That said, it does not alter the fact that household chemicals are a source of worry to all parents, and if things were made safer a lot of valuable medical time could be saved, and the child would be less likely to be traumatized by the event. Nor should we forget the plight of the parents, who are often more upset than the child.

An obvious precaution to take is to make household chemicals taste repellent, and what is needed is a substance that is extremely bitter, this being the taste sensation we like the least. (In Victorian times, when the threat of accidental poisoning came often from family medicines, these were made to taste very bitter so that children would not consume them.) Yet the additive must not interfere with the function for which the household chemical is required, and only a tiny amount of the bittering agent should be needed, so that manufacturers are not deterred from adding it for economic reasons.

Nature produces several very bitter substances, such as bitter aloes, wormwood from the leaves of *Artemisia absinthium*, gentian from the root of the *Gentiana lutea*, and quassia from the stem of the *Jamaica quassia*. These have sometimes been used to deter drinkers, but often they have been used for the opposite effect. Added to aperitifs they are thought to stimulate the appetite. But the molecule that out-tastes all others in the bitterness stakes is Bitrex. Bitrex was discovered in 1958 at the Scottish pharmaceutical company, T. & H. Smith Ltd of Edinburgh, during a search for new lignocaine painkillers of the type that numb the skin when applied to the surface. Bitrex is a non-toxic white powder which is soluble in all kinds of solvents and is even listed in the *Guinness Book of Records* as the bitterest substance known. We can detect Bitrex at the level of only 10 ppb and register it as bitter at 50 ppb. It is generally used at levels of parts per million, for example industrial alcohol is denatured with 10 ppm Bitrex.

Bitrex is a perfect fit for the bitter taste receptors on our tongue, which explains why it provokes an immediate response, but it lingers long once you have taken some into your mouth. Bitrex's chemical name is denatonium benzoate, and it is the denatonium part which is the active component. This has two ethyl groups and a benzyl group attached to a nitrogen atom. If these groups are replaced by the chemically very similar, but smaller, methyl then the bitterness is reduced 100 000 times. Indeed, any slight changes to the Bitrex molecule make it much less effective.

As well as being added to household chemicals, Bitrex has found many other uses. Products that could contain Bitrex as a protective are: polishes, air fresheners, hair colourings, medical wipes, rubbing alcohol, car washes, chrome polishes, firelighters. Paint containing Bitrex can be used to coat old lead-containing paintwork and so discourage young children from picking off flakes and eating them, which some are prone to do. A few manufacturers resist using it, although the tiny additional cost this adds cannot really be the reason. It has been put in slug pellets to deter birds, on young trees to stop animals feeding on their leaves and bark, and in special nail varnish to deter nail biting. The biggest use of Bitrex is to denature industrial alcohol.

Portrait 10

Elements from heaven (1)—zirconium

You may be reading this with zirconium hydroxychloride under your arms, especially if you used an anti-perspirant this morning. This curious compound is made from a little known metal which is of growing importance.

The year 1789 is best known for the start of the French Revolution. But there was also another event that occurred in Germany that year, and one that is set to revolutionize our lives in the next century. That event was the discovery of the element zirconium, one of the safest of the chemical elements. Products made from zirconium will be coming into our homes in some bizarre disguises, such as ceramic cutlery, fake diamonds and new colours. Some are already with us. Industry too has found remarkable uses for this metal and especially for its oxide, called zirconia, which melts only above 2500 Celsius.

The chemist Martin Heinrich Klaproth (1743–1817) discovered zirconium in semi-precious stones from Ceylon. He had trained as an apothecary, but he became interested in analytical chemistry and was eventually appointed the first professor of chemistry at the University of

Berlin when he was 60. In the same year that he discovered zirconium, he also discovered uranium. The fates of these two elements were to intertwine, as we shall learn.

Gems that contain zirconium were known in biblical times, and called by various names such as hyacinth, jacinth, jargon and zircon. Colourless varieties were thought to be an inferior kind of diamond. This was shown to be false when Klaproth decomposed a zircon gemstone by heating it with alkali. From the product of that reaction he extracted a new oxide which he called zirconia, and which he realized was of a new element, zirconium. The name comes from the Arabic word *zargum*, meaning gold-coloured.

Klaproth was unable to isolate the pure metal itself, and sadly he did not live to see this achievement, which was accomplished in 1824 by the Swedish chemist Jöns Jacob Berzelius. Even so, for the next 120 years little use was found for zirconium. It had no commercial application as a metal, and its chemical compounds were without any noteworthy features. Yet zirconium metal had a hidden asset which suddenly brought it to prominence in the 1940s, when atomic energy was first liberated. Zirconium proved to be an ideal metal for the inside of nuclear reactors. It does not corrode at high temperatures, nor does it absorb the neutrons which the reactor produces, a process which would turn other metals into dangerous radioactive isotopes. The nuclear industry is still the major end-user of the 7000 tons of zirconium metal smelted each year.

Zirconium is not a particularly rare element. It is almost three times as abundant as copper and ten times as abundant as lead, and there is 80 times more zirconium than tin in the Earth's crust. Australia, South Africa, India and the USA have vast deposits of zircon (which is zirconium silicate) and of zirconia (which is zirconium dioxide). Global production of pure zirconia is now almost 25 000 tons per year, and it is used for making everything from anti-perspirants to brilliant gemstones, called zircons.

Happily, zirconium poses no biological threat, either to human health or to the environment. Indeed, manufacturers are changing over to zirconium chemicals as safer alternatives. Even the tiny amounts of lead compounds still added to some paints, to improve drying, are being replaced by zirconium salts. The paper and packaging industry is finding that zirconium compounds make good surface coatings because they have excellent water resistance and strength. Equally important is their low toxicity, and a zirconium carbonate compound is approved for treating wrapping paper that comes in contact with food. Zirconium hydroxychloride is now the preferred anti-perspirant component in roll-on deodorants, replacing the aluminium compounds previously used.

Over 600 000 tons of zircon sand are mined each year to be used as such in heat resistant linings for furnaces and to make giant ladles for molten metal. It is impervious to high temperatures and does not expand when heated. However, the most dramatic use of zirconium is as its dioxide in super-strong ceramics. These were developed as a way of making engines for military tanks that would need neither lubricating oil nor a cooling system. From this research came a new generation of tough, heat resistant ceramics that are stronger than metals. In Japan these are used for high speed cutting tools for industry, and for knives, scissors and golf clubs for the domestic market. Zirconia also finds its way into TV screens, where it protects us from harmful X-rays.

Portrait 11

Elements from heaven (2)—titanium

Two years after the discovery of zirconium, another metal very similar to it was discovered: titanium. This find was rather unusual, as it was made by a clergyman, the Reverend William Gregor (1761–1817), at a remote village in Cornwall, England. Today this metal is found in aircraft engines, in deep-sea oil rigs, on ships, and in the elbow of the heir to the British throne. Titanium dioxide is the world's most important pigment, and is the brilliant white of kitchen appliances, plastic plumbing, household paints, and the white paint used for lane markings on roads.

About 200 years ago the Reverend Gregor, who was the vicar of Creed, analysed a black sand that he had found in the nearby parish of Menaccan. He found the sand odd because it was attracted to a magnet. He analysed it as best he could and deduced that it was made up of two metal oxides—iron oxide was one, but the other he could not identify. He was enough of a scientist to realize that it must be the oxide of an unknown metal, and he reported it to the Royal Geological Society of Cornwall. He also wrote a paper about it, which appeared in the 1791 edition of the German science journal *Crell's Annalen*. Four years later Klaproth, the man who had come across the new metal oxide zirconia, now 'discovered' another oxide of an unknown metal element, and he gave it the name titanium.

Titanium is the seventh most abundant metal of the Earth's crust. Neither Gregor nor Klaproth lived to see the metal itself, which is very difficult to extract from its oxide ores. Impure samples of titanium were made in the nineteenth century, but it was not until 1910 that Matthew Hunter, working for General Electric in the USA, made absolutely pure titanium. The titanium industry really started in the 1930s when paint manufacturers were seeking a replacement for white lead and turned to titanium dioxide. Just as the oxide of zirconium is called zirconia, so the oxide of titanium is called titania. This also happens to be the name of the Queen of the fairies in Shakespeare's play *A Midsummer Night's Dream*, which perhaps explains why those who work in this industry prefer the name titanium dioxide, or the chemical formula TiO_2. This chemical is now a three million tons-a-year industry. Titanium dioxide became the most commonly used pigment because it is non-toxic, does not discolour and has a very high refractive index. The refractive index is a measure of a substance's ability to scatter light, and explains the brilliant whiteness titanium dioxide imparts to domestic appliances such as fridges, washing machines and driers. Its refractive index of 2.7 is even greater than that of diamond (2.4), which is famed for its brilliance.

Half of the titanium dioxide output ends up in paint, a quarter goes into plastics such as carrier bags, windows and pipes, and the rest goes into paper, synthetic fibres and ceramics. A little is used in cosmetics, and it is even safe enough for use in foods such as icing sugar and sweets. Titanium dioxide also has the ability to absorb the damaging UV rays of sunlight, which is why it is used to protect the plastic of window frames and to protect the human body in sunscreen ointments. The world's largest producer of titanium dioxide is Du Pont of the USA, while in Europe the biggest producer is Tioxide of the UK.

Titanium dioxide is made by two processes. The older one involves dissolving titanium ore in sulphuric acid, precipitating the wet oxide and heating it at 100 Celsius. The more modern process uses chlorine gas to convert the ore to titanium tetrachloride, which is then oxidized with oxygen at 1000 Celsius, or, better still, in a plasma arc at 2000 Celsius. Titanium tetrachloride is the key chemical in the titanium industry. This crystal clear, volatile liquid boils at 136 Celsius and is consequently easy to purify. Heating it with either magnesium or sodium metal in an electric furnace releases titanium metal.

Although it is difficult to produce, titanium justifies its cost. The metal is strong, has a high melting point of 1660 Celsius (slightly higher than that of iron) and is light, and so it is used in aircraft engines and airframes alloyed with small amounts of aluminium and vanadium. Titanium is very reactive, but it is protected by a thin oxide layer on its surface. This enables it to resist the corroding action not only of sea-water and bleach, but also of corrosive chemicals. Some Russian submarines have titanium hulls. The metal will also withstand nitric acid and chlorine gas, which is why chemical engineers rely on it for chemical plants; while mechanical engineers find it ideal for heat exchangers in power stations. It is much used in off-shore oil rigs. Around 50 000 tons of titanium metal are produced each year.

In the 1950s surgeons noted that titanium was ideal for repairing broken bones. It is unreactive, so is not corroded by bodily fluids; it is non-toxic, bonds to the bone, and is not rejected by the body. Hip and knee replacements, pace-makers, bone-plates and screws and cranial plates for skull fractures are made of titanium and can remain in place for up to twenty years. Prince Charles had his broken elbow repaired with a titanium plate by surgeon John Webb at Nottingham University hospital. Barry Sheen, the motor cycle racer, is reputed to be held together by titanium supports following a terrible accident which shattered many of his bones.

Titanium implants are used for attaching false teeth. Plugs of the metal are inserted into the jaw bones using a technique developed by Per-Ingvar Brånemark of Göteborg, Sweden, who has patients with implants he inserted as long ago as 1965. The key to success with titanium implants is pure titanium and scrupulous cleaning. For this purpose a plasma arc is used which strips off the surface atoms and exposes a new layer of the metal that is instantly oxidized. It is to this oxide film that body tissue will bond so strongly.

Nothing now remains of the sand by the river bank along which the Reverend Gregor first came across a titanium ore. The site was built over long ago. Titanium ore is not mined in Cornwall where it was first found, but vast deposits in North America, South Africa and Australia supply the world's needs. South Africa has coastal sand dunes 180 kilometres north of Durban, where the waves of the Indian Ocean have deposited the minerals rutile, zircon and ilmenite over millions of years. Rutile is titanium dioxide, zircon is zirconium silicate, and ilmenite is a mixed oxide of iron and titanium. A dredger floats on a large artificial pond and sucks up the sand which contains about 5% of ilmenite, rutile and zircon. The ilmenite is separated, upgraded and smelted to produce titanium dioxide slag and pig iron. The rutile and zircon are separated in a dry electrostatic process, and are sold in their natural form. As the dredger moves on it is followed by an environmental rehabilitation programme which recreates flourishing woodlands, wetlands and grasslands.

GALLERY 5

MATERIAL PROGRESS AND IMMATERIAL OBSERVATIONS

An exhibition of molecules that make life a little easier

■ BACK TO THE FUTURE ■ PLASTIC DICKIES AND EXPLODING BALLS ■ BEND ME, SHAPE ME, ANY WAY YOU WANT ME ■ CHEAP AND CHEERFUL ■ GOING TO EXTREMES ON EARTH AND IN THE HEAVENS ■ GETTING RID OF UNWANTED PETS ■ SEXY AND SAFE ■ FILM SET FUN AND FAST FOOD FLAW ■ STRONGER THAN STEEL

YOU MAY think of polymers as entirely manufactured and therefore unnatural, but they are often the chemists' attempts to supplement and improve on the biological polymers that nature produces. Cotton, ivory, leather, linen, paper, rubber, silk, wood and wool are wonderful materials made from the biological polymers that plants and animals produce, and which have evolved to serve such useful ends as providing protective outer layers, insulation, reinforcement, weaponry and so on. Humans learned that with a little modification they could turn these polymers into quite useful articles, such as briefs and briefcases, condoms and tea cosies, tickets and toothpicks.

Sometimes we want polymers with features that never evolved in nature, such as non-cracking insulation for electric cable, clothes that can be unpacked after a long voyage and still be without creases, or pans in which to fry eggs without them sticking. For these polymers we have had to look to chemists. Most of the portraits in this Gallery are of these kinds of polymers—materials that do not have natural equivalents.

Polymers are rather special kinds of molecules consisting of long chains, usually made up of carbon atoms, to which other atoms, such as hydrogen, fluorine and chlorine, are attached. The older name for polymers is plastics, and you probably know several of them by name—polythene, polystyrene, Teflon, Orlon—but these are only a few of the many that now play an important role in our lives. Whatever role polymers play, they cause many of us to adopt quite strong attitudes towards them. A few of us admire them, many of us ignore them, but a growing number despise them and a few abhor them and will avoid them at all costs. To a chemist, this opposition to polymers seems rather strange. By the time you come to the end of this exhibition I hope that visitors with strong views will have seen enough to persuade them to change their mind.

Attitudes towards plastics have changed over the past half-century. In the 1930s, when cellophane, PVC, polystyrene, Perspex and nylon were launched, plastics were welcomed. This was even termed the Plastics Age, and there was widespread approval, particularly from young and influential designers who welcomed the new materials that chemistry had produced, and that seemed impervious to rot and decay.

The contribution of the new plastics to the Allied victory in 1945 ensured a continued expectation of wonder from these new materials, and the 1950s and 1960s saw the chemical industry delivering some remarkable products which transformed textiles and home furnishings. Over-confidence was bound to lead to mistakes, and plastic made three gaffes in the 1950s. It appeared in the guise of plastic flowers, throwaway cups and cutlery, and plastic film used to wrap dry-cleaned clothes.

The first was merely an aberration of taste, but the second emphasized plastic's intrinsic worthlessness, and the third was a disaster because it could cause babies to suffocate. When plastics attracted the wrath of Norman Mailer then their humiliation was assured, and from the mid-1960s onwards he ran a hostile campaign against them. 'We divorced ourselves from the materials of the earth, the rock, the wood, the iron ore,' he maintained. 'We looked to new materials which were cooked in vats, long complex derivatives of urine which we called plastic.' The new materials lacked 'the odour of the living,' their 'touch was alien to nature,' and they proliferated 'like the metastases of cancer cells.'

With such vitriolic remarks from one of the world's leading playwrights, it hardly needed the environmentalists to put the boot in, but they did. Plastics seemed to symbolize the wasting of the Earth's resources and the polluting of the planet. Suddenly the permanence of plastic was not an asset but a major fault. And yet the portraits in this Gallery are not a collection of villains and monsters, but of skilled designers, honest labourers, space explorers and miracle workers.

Portrait 1

Back to the future—Tencel

Not surprisingly, chemists are sometimes unable to better a polymer that nature has produced. Plants produce one of these superbly well: cellulose. This is made from the carbohydrate, glucose, which in turn is made from water and the carbon dioxide in the air. These are induced to combine by tapping the energy of sunlight in a process called photosynthesis, which occurs with the help of the green catalyst chlorophyll. This primitive chemical reaction generates the glucose molecule which is the most abundant biological molecule produced on Earth, to the extent of around 50 billion tons a year. It occurs mainly as the polymers starch and cellulose.

Glucose can link itself together in long chains of two kinds: the first is one which can easily be broken down again into its constituent units, and we know this as starch; the second is one which is almost indestructible and can last for thousands of years, and we know this as cellulose. Starch is a way for the plant to store food, and cellulose is the way it builds its more permanent structures, such as roots, stems and leaves. The difference between starch and cellulose is simply in the way in which the glucose molecules are joined to one another; starch can be easily unhooked by digestive enzymes, but cellulose can not.

Our primate ancestors soon learnt the difference between those parts of a plant which were rich in edible starch, such as seeds, fruit and tubers, and the bits that were mainly tough and indigestible cellulose such as stalks, leaves and husks, although even some of this can serve a useful purpose as fibre in our diet. Cellulose can be used in other ways, and when it is in a fibrous form we can spin and weave it, as with cotton or flax. And while the cellulose of wood is not easily accessible because it is intermingled with non-fibrous material, it can nevertheless be turned into an equally valuable commodity, paper. Yet the cellulose of wood is no different from that of cotton, and with a little skill it too can clothe us and furnish our homes.

Cellulose will one day be the raw material of a large part of our chemical industry, and as the drive to renewable resources continues then this may well happen in the next century. In which case, it is interesting to examine the portrait of cellulose to see what it holds in store, as a fibre for making fabrics, and as a raw material for making plastics. Both uses were developed in the nineteenth century, when artificial silk and celluloid were invented. One of these has recently been upgraded with new chemical technology; the other is probably best left in limbo, as we shall see.

What led chemists to seek a new fibre was the popularity and expense of an existing one: silk. Silk has a softness, a feel and a rustle that makes it very appealing. Perhaps it should always be an exclusive fibre, verging on the decadent in the form of silk pyjamas and silk sheets, whose sensuousness has long been used to emphasize the ultimate in luxury fabrics. But in Victorian times silk became associated with mourning, especially for widows. The death of a loved one meant dressing elegantly in black, but there was a limit to what the silk worms of China could produce for the sad widows of Western society.

The first artificial cellulose fibre was marketed by the French chemist, Comte de Chardonnet, who introduced a nitrocellulose cloth version in 1884. To begin with it sold well, and production of so-called Chardonnet silk soon reached 10 000 tons per year. Unfortunately it had an upsetting tendency to burst into flames, and sometimes it even exploded (see Portrait 2). A safer cellulose fibre was developed by Charles Cross, Edward Bevan and Clayton Beadle, working at a small shop in South Avenue, Kew, on the outskirts of London. They made the first commercially successful rayon and patented it in 1894. Their new material, art silk, was just as good as natural silk and cost a fraction of the price. Art silk is short for artificial silk, although that name is rarely used today, and it is better known under its other names of rayon, acetate and viscose. Cross, Bevan and Beadle sold their invention to the Courtaulds, who were the main silk producers at the time, and they began manufacturing the new fibre in 1905, although they had to overcome the tricky problem of extruding it through tiny nozzles to get the fineness they sought.

Three million tons of rayon are made each year, but this has been a declining fraction of the world market in artificial fibres. Rayon once reigned supreme before the advent of nylon in the late 1930s, and it still has many advantages: it feels gentle to the skin, it hangs nicely and it dyes well. Rayon is widely used for underclothes, blouses, dresses and jacket linings, and it is blended with other fibres in bed linen, upholstery and curtains. But it has its drawbacks: it creases easily, it lacks strength when wet—and it can lose battles.

In their attempt to create an autarkic state, the Nazis invested heavily in rayon, and production rose ten-fold from 1933 to 1943. An autarkic state is one which tries to operate as a self-sufficient economic system, and is much beloved by dictators, because it means they cannot be affected by international trade sanctions or boycotts. The Nazi state decided to make itself independent of imported cotton by using rayon instead, which it could manufacture from the abundant forests in its territories. German soldiers were kitted out in uniforms made of rayon, and this is sometimes given as the reason why so many fell victim to frostbite on the Russian front—it may have contributed to their defeat at the battle of Stalingrad in the winter of 1942/3. Even though rayon is probably not the best material for military uniforms, the Nazi scientists made a superb type of fibre, and developed a continuous process of manufacture which was taken over by the Allies at the end of the War to become the standard method of production throughout the world.

Rayon is made from the cellulose of wood pulp, and to convert this plant material into fibres for textiles it is necessary to solubilize it. This is not easy, because the chains of cellulose polymer cling strongly together. The only way to untangle them and pull them out into fine threads is to dissolve them. For rayon this is done with caustic soda solution and carbon disulphide. When the viscous solution that ensues is forced through tiny nozzles it forms fibres of rayon, and when it is forced through a narrow slit it emerges as a sheet of cellophane. At the same time the solution is neutralized with acid to make the cellulose insoluble again. The process is efficient, but produces a lot of foul-smelling effluent.

For 90 years the chemistry of the process changed little, but about 15 years ago Courtaulds' research chemists chanced upon the curious discovery that cellulose would dissolve very well when heated in the solvent N-methyl morpholine oxide. This is the basis for the production of their new fibre Tencel. To turn the resulting viscous solution into the new fibre required a new technology of solvent spinning. The fibre is extruded into air and immediately passed into water to wash out the solvent, which is reclaimed for re-use. The result is that there is almost no chemical effluent from a Tencel factory.

Tencel was developed by chemists at Courtaulds' research labs in Coventry, England. Courtaulds has built new plants, the first at Mobile, Alabama, USA, and the second at Grimsby, on the north-east coast of Britain, where production should reach 100 000 tons a year by 2000. Courtaulds is very proud of Tencel: it is the first new fibre to be launched in 30 years; and it is very different from the older forms of rayon. It has a luxurious feel and a fluid drape, yet it can be used to make robust clothes like jeans and chambray shirts. The new fibre also has better wet strength, lower shrinkage and is crease-resistant.

It is not the technology, nor the environmental benefits, that sell Tencel, but its quality. Courtaulds are keeping it exclusive by deliberately pricing it high, and people are willing to pay the extra for it, as its impact in Japan has demonstrated: there, Tencel jeans have been sold at many times the price of ordinary denim jeans.

Portrait 2

Plastic dickies and exploding balls—celluloid

Modern plastics are so strong, so versatile, so safe, so dull—which may be why the public finds them rather boring. How different was the first successful plastic, celluloid. Rayon may have been redeemed as Tencel; but celluloid, the old type of plastic which was made from cellulose, will probably never return to its former popularity, though it is still has its niche uses. It is mainly cellulose nitrate, and though it is highly flammable, it is still used to make table tennis balls and nail varnish, which are rarely a cause for alarm.

People were fascinated by celluloid when it first appeared in the 1860s, and by the many guises in which they encountered it, as buttons and billiard balls, trinket boxes and toys, collars and dickies—those uncontrollable false shirt fronts that featured in early comedy films. Celluloid was welcomed as part of a new age of chemistry. It was cheaper than ivory, the material it was mainly replacing, and demand for which could only be satisfied by the slaughter of 20 000 elephants a year. Admittedly celluloid was dangerously flammable, and at times it could even be explosive. There were tales of celluloid billiard balls colliding so violently that they went off with a bang. When this happened in a saloon in Colorado it started a gunfight . . . or so said John Wesley Hyatt, the man who popularized celluloid, in his 1914 acceptance speech for the Perkin Medal, which he was awarded by the London Chemical Society, now the Royal Society of Chemistry. By then celluloid had been on the market for almost 50 years, and many thought Hyatt had even invented it.

The story of celluloid began in 1845, when the Swiss German chemist Christian Schönbein reacted cotton with a mixture of concentrated sulphuric acid and nitric acid. The resulting material was cellulose nitrate, and it had a rather remarkable property. When compressed into blocks it was extremely explosive, and this became one of its major uses as gun cotton or nitrocellulose.

Depending on the conditions and the extent of nitration, the product could vary from a plasticine-like solid to a viscous liquid. The explosive version had three nitrate groups attached to each glucose ring, but if nitration was limited to two groups, then the product did not explode, although it was still highly flammable. This form of cellulose nitrate reached the public in two forms: collodion and celluloid. Collodion was a solution of cellulose nitrate dissolved in a 50/50 mixture of alcohol and ether (these solvents are more correctly called ethanol and diethyl ether), and was the basis of fast-drying enamel paints. It was eagerly bought by youngsters as an adhesive for model building, while their parents kept a small bottle in the medicine cabinet and used it to remove painful corns from their toes. Celluloid was a plastic version of cellulose nitrate formulated with about 20% camphor, which is a waxy solid obtained originally by distilling the bark of the camphor tree which grows in Japan. Camphor is a cyclic molecule which melts at 179 Celsius; it can also be made synthetically. The blend of cellulose nitrate and camphor gave a dough that could be coloured, softened by heating, and pressed into moulds. The camphor also made it less flammable.

According to Susan Mossman of the London Science Museum, in the book *Development of Plastics*, it was not Hyatt who was the first to make celluloid, but a Briton, Alexander Parkes. He modestly called his new plastic Parkesine, and exhibited it at the second Great Exhibition in London in 1862. He set up a company to make celluloid products, but they proved much inferior to their ivory counterparts and the company soon went bust. Daniel Spill, who had been Parkes' works manager, launched his version of celluloid, Xylonite, in 1869, but his business also failed.

Then in 1870, across the Atlantic, Hyatt started his firm, making what he called Celluloid, a name which stuck because his product was a success. Soon it was being turned into spectacle frames, false teeth, piano keys, plastic madonnas, cosmetic jars, cutlery handles and dog-collars for clergymen. Admittedly they burned well, but they didn't explode. So why did the celluloid billiard balls occasionally explode? This was probably caused by painting them with collodion to give them a hard and shiny finish, leaving a film of pure cellulose nitrate on their surface. This in itself might not have been thick enough to explode when two balls collided violently, but it might have been enough to cause part of the underlying material to go off with a bang, particularly if this had been unevenly mixed with the camphor.

Portrait 3

Bend me, shape me, any way you want me—ethylene

The petrochemical industry produces the gases ethylene and propylene in vast amounts. They are turned into hundreds of other chemicals, and in particular into polyethylene (popularly called polythene) and polypropylene. These workhorse plastics have been around for longer than half a century, but as with rayon, a remarkable breakthrough has recently been made, thanks to new catalysts, leading to new forms of these polymers with new uses.

Ethylene, more correctly called ethene, is a gas at normal temperatures, condensing to a liquid at −104 Celsius. It can be transported and stored as such. It is a small molecule, formula C_2H_4, with two carbon atoms joined by a double bond. It is this double bond that makes ethylene chemically reactive, and able to form polymers. Ethylene is the gas which causes blossoms to fall—in nature, it is a plant hormone; it also causes economies to bloom: it heads the list of organic chemical production. Worldwide, the amount of ethylene produced has risen from 2 million tons a year in 1960 to 67 million tons in 1995, and is expected to reach 100 million tons by the year 2005. Despite this, ethylene can be unprofitable to make because of over-production, and yet we cannot stop making it because from it come many of the benefits of civilized living in our homes, shops and offices.

While ethylene is a feedstock for the chemical industry, it also has a key role in nature. In the 1930s, biochemists realized that ethylene was produced during many stage of a plant's lifecycle—germination, growth, blooming, fruit ripening, ageing, leaf loss—and in response to damage caused by extremes of temperature and drought. The concentration of ethylene in plants is normally low, but can reach as much as 2000 ppm (0.2%) in ripe fruit. Plants make their ethylene from methionine. This amino acid converts to a cyclopropane derivative with a highly strained ring of three carbon atoms. When this comes in contact with an oxygen molecule of the air it breaks apart and expels two of its carbons as ethylene.

In the nineteenth century people were often puzzled by the odd behaviour of trees planted along city streets. For no apparent reason they would shed their leaves in the middle of summer. This curious phenomenon was eventually traced to leaking gas mains, and in particular to the ethylene in the gas which was at a level of around 10 ppm. This was also the reason why our great-grandmothers would avoid putting plants and flowers in a room with a gas fire because they believed this would cause the leaves or petals to fall. Perhaps they also knew that if this happened, then they had a slight leak in their gas piping.

For 50 years, fruit importers in temperate countries have been copying nature, using ethylene to ensure a regular supply of ripe tropical fruits such as bananas, avocados, nectarines and peaches for northern markets. The fruit is picked unripe, transported to its destination, and ripened as required by giving it a whiff of ethylene gas. This is generated either by passing alcohol vapour over an electrically heated catalyst, or by dissolving the chemical chloroethylphosphonic acid in water. Fruit starts to ripen when ethylene in the surrounding air reaches just 1 ppm. We can encourage ripening in our own home, which we might want to do if we have grown tomatoes outdoors and had to pick them while they were still green because of an impending frost. The trick is to put them in a dish with a ripe banana: the banana gives off ethylene, and the tomatoes start to ripen. In some countries ethylene is actually generated artificially in orchards and fields to ensure uniform ripening of figs, mangoes and melons, and to induce blossoming of pineapples, and to make olives easier to pick. Understanding how ethylene works as a plant hormone has also enabled biotechnologists to redesign the humble tomato, producing a new variety which ripens more slowly and stays firmer longer, and so can be transported over larger distances.

It is possible to protect fruit and flowers against ethylene, and so extend their lives. Fruit can be stored in sealed containers with sachets of silica which is impregnated with potassium permanganate. This chemical mops up ethylene. Flowers can be wrapped in plastic film which is impregnated with a similar ethylene-absorbing material.

While plants can make ethylene at ordinary temperatures, industry has to use high temperatures to generate the gas on the scale which the world's economies need. Ethylene can be produced by heating hydrocarbons with steam at temperatures in excess of 800 Celsius. The total global ethylene capacity is around 75 million tons, of which the USA accounts for a third, Western Europe a quarter and Japan a tenth. More recently, South Korea has invested in ethylene production and has expanded this fivefold in five years to around 3 million tons, far in excess of local needs. They are clearly hoping to supply China with ethylene in the next century. Pipeline networks exist in North America and Europe to move ethylene around and some are linked to massive underground storage caverns capable of holding 3 million tons of the gas.

Ethylene is a measure of a country's economic strength. One hundred and fifty years ago, the German chemist Justus von Liebig said that sulfuric acid production was the most accurate guide to a country's industrial production, and while this chemical is still produced on a scale which dwarfs all others, it has passed its role as an economic indicator to ethylene. In the developed industrialized nations demand for this industrial gas ebbs and flows in phase with the trade cycle, while in other parts of the world there is still a steadily growing demand. It is this which explains the overall annual global growth rate of around 4% per year, which is predicted to continue well into the next century.

Despite all this, ethylene has rarely been a profitable commodity for the primary producers. While older plants have been closed, new capacity has more than kept pace with rising demand and plant closures. Profitable or not, a modern industrial economy cannot do without ethylene and the products made from it. Around half is turned into polyethylene for packaging, pipes and sheeting, while the other half ends up in guises as diverse as window frames and water mains, perfumes and painkillers. Before it does it has to be turned into intermediates such as ethylene dichloride, ethylbenzene, ethylene oxide and ethylene glycol, all of which are important in the world economy. Ethylene is also turned into ethanol, acetaldehyde, ethyl chloride, ethylene dibromide and acetic acid, and these reach the consumer in products as diverse as garden seats and tasty treats such as salt and vinegar crisps.

Polyethylene was discovered in 1933 by Reginald Gibson and Eric Fawcett at the chemical company, ICI, at Winnington in the UK. Under the right conditions molecules of ethylene link up to form the chains that are responsible for the properties that make polyethylene ideal for buckets and bowls, plastic bags and sacks.

Low density polyethylene (LDPE) is made at high pressure and ends up mainly as film and packaging, while high density polyethylene (HDPE) is made at lower pressures and is used to make containers and pipes. Yet even these long-established work-horse materials are being challenged. LDPE will be replaced by LLDPE (the extra 'L' stands for linear) thanks to new and better catalysts, the metallocenes. These are compounds of titanium, zirconium or hafnium, in which the metal atom is sandwiched between two rings of carbon atoms. The metal is thereby kept in an electron-rich environment and is held in such a way that it not only behaves as a catalyst, but also directs the way in which the polymer is formed. We shall see the importance of this in the next portrait, which is of polypropylene.

The length of the polymer chain which forms when metallocenes are used depends very much on the temperature at which the polymerization is carried out. The lower the temperature, the longer the chain. Most common polyethylene grades have chains ranging from 1500 carbon atoms to 20,000. At 20 Celsius zirconocene catalyst gives a chain of 50 000, while at 100 Celsius the chain is less than 1000 atoms long. Between these temperatures the process can be finely tuned to give just the optimum length for the intended application. The length of a polymer chain has a marked effect on properties such as its softening temperature, flexibility and toughness. Because of this, polyethylene is about to experience a renaissance and even invade other polymer markets, while copolymers are producing even more exciting variants. Copolymers are plastics that are made by polymerizing two molecules at the same time so that the product has features of both. It has been estimated that by the year 2000 metallocene-derived polymers could well account for 20 million tons, or 10%, of the world market.

Modern life would be unthinkable without the products derived from ethylene, so how will future generations fare if the world decides to stop drawing on its fossil reserves? Could all the polyethylene we need be produced from renewable resources? The answer is yes, and the source would be alcohol (ethanol). Crops such as sugar cane and cereals can be turned into ethanol, and this can be turned into ethylene, and then into polyethylene.

Wonderful as polyethylene is, this polymer is still not perfect. The trouble with polythene bottles is that although they hold water, they tend to soften and even dissolve into holes when some liquids are put in them. This problem can be solved by coating the surface of polythene with a thin layer of a tougher plastic. This layer is simply made by exposing the polythene container to fluorine gas. Such is the strength of the new super-polythene that it is even being used for petrol tanks in cars. It will cope with liquids such as oil, cleaning fluids, printing inks, cosmetics, toiletries, and even the most concentrated sulphuric acid. It is also suitable for food stuffs, like cola concentrate.

The success of super-polythene lies with the toxic gas, fluorine. When the surface of ordinary polythene is exposed to this, it reacts to form another polymer, polyfluoroethylene. The carbon–hydrogen bonds of the original polythene have been swapped for carbon–fluorine bonds which are much stronger. In effect the surface has become coated with the same kind of fluoropolymer that is used on non-stick frying pans. (Portraits of other members of this family of polymers are also on display here, one later in this Gallery (portrait 5) and the other in Gallery 4 (portrait 3).) This barrier layer of fluoropolymer is less than a hundredth of a millimetre thick, and just how thick depends on the time the polythene is exposed to the fluorine. The fluoropolymer layer gives the container two advantages: it makes the surface resistant to all forms of chemical attack, and it repels all other liquids. Nor does it mask the colour of the underlying material.

Taming fluorine gas so that it can be used as an industrial chemical is the secret of the process. The fluorine is made by passing an electric current through molten potassium bifluoride. It was first made in this way by Henri Moissan in 1886, and he won the Nobel Prize for chemistry in 1906 for his work. Fluorine is the most reactive of all the elements; it will even cause iron wool to burst into flames. By diluting it with nine parts of nitrogen gas its raw power can be controlled, and this is how it is shipped and transported. For use in fluorinating polythene it may be diluted with even more nitrogen.

The surface layer of fluoroethylene can be produced in two ways. The fluorine/nitrogen gas is itself used to blow the molten polythene into moulds for bottles and other containers. Within seconds the fluorine forms an inner layer of fluoropolymer on the polythene. A better method is to carry out the fluorination as a separate step, and this has the advantage that both the inside and outside of each container are coated with fluoropolymer, and to whatever thickness of barrier the customer desires. Surface fluorination adds only a few pence or cents to the price of a litre polythene bottle—a cost that can be easily absorbed whether the container is as tiny as a mascara jar or as large as a thousand-litre tank.

A gas as dangerous as fluorine naturally raises problems of health and safety. Breathing fluorine gas at concentrations of 0.1% for only a few minutes will kill. Luckily the gas is so pungent that it acts as its own early warning system, but in any case stringent legal controls surround its use.

Lethal as fluorine is, when put to use in strengthening polythene it may end up saving lives. Petrol tanks made of plastic are less likely to rupture in a crash, and so less likely to cause a fire. Ordinary polythene may have the right resilience and strength for a fuel tank, but unfortunately it lets fuels escape slowly through its walls. For example, diesel fuel will evaporate through polythene at the rate of 2% per week. Super-polythene, on the other hand, prevents this.

Following its success with polythene, we may eventually see the surface fluorination of other plastics and polymers. Surgical gloves treated on the inside with fluorine require no talcum powder on the hands since the fluoropolymer acts as its own lubricant. Cables sheathed in polythene can be made even more weather resistant. The same is true of the blades of windshield wipers. Indeed, any rubber that is exposed to sunlight and ozone, which cause brittleness and cracking, could be improved by fluorination.

And as for recycling, we need not worry that this surface layer of super-polythene will interfere. Unlike some polymers, which cannot be recycled if a small amount of another, incompatible, polymer gets mixed in with it, polyethylene is not affected at all by fluorinated polyethylene. On melting, the surface layer of fluoropolymer simply becomes incorporated into the main body of the plastic and produces a slightly stronger material for its second time around.

Portrait 4

Cheap and cheerful—polypropylene

Propylene is very similar to ethylene: it is the same core molecule with the same double bond, but has an extra carbon atom in the form of a methyl group (CH_3) attached. It polymerizes in the same way as ethylene. However, the product, polypropylene, despite its own benefits, seems always to have been in the shade of its more famous sister, polyethylene.

Polymers, like movie stars, need a good name if they are to become a big success. Unfortunately polypropylene never rated a star name, like polythene, nylon, Teflon and rayon. This is a pity, because polypropylene brightens our lives in items as diverse as ornamental rugs, kitchen kettles, shatterproof glasses, colourful crates, wrappers for chocolate bars, artificial turf, garden chairs, rugged suitcases, car bumpers, margarine tubs, CD cases, towing ropes, baling twine—and tea-bags. Cars in particular contain a lot of polypropylene, in the form of dashboards, bumpers (fenders), battery cases, upholstery and carpets. Some cars contain as much as 80 kg of this wonderful material. Part of its appeal is that we like the feel of polypropylene against our skin, and it has the benefit of allowing moisture to escape while keeping us dry. This is why it is used in thermal vests, disposable nappies and even space suits for astronauts.

Polypropylene can range from being a rigid plastic to a fluffy fibre; it can be crystal clear or multi-coloured; as soft as silk or as hard as iron. Polypropylene is heat stable and sterilizable, impermeable yet flexible, transparent and perfectly safe. Yet even a plastic as versatile as polypropylene can be radically transformed and its range of uses greatly expanded, as we shall see.

Polypropylene was first made in 1951 by two chemists working for Phillips Petroleum in the USA. As they watched the gas propylene turn into a toffee-like solid they knew that they had stumbled onto something remarkable. The company patented the process, and the discoverers, 32-year-old Paul Hogan and 30-year-old Robert Banks, worked together for many years developing the product. Recognition for their achievement, however, had to wait 36 years, when they were honoured with the prestigious Perkin Medal by the UK Society of Chemical Industry. The delay was partly caused by the fact that half a dozen other companies also filed patents on polypropylene, and the legal battles only ended in 1982. Since then, polypropylene has burgeoned. It is now regarded as the barometer of GDP for industrial economies. Global capacity for polypropylene in the mid 1990s was more than 20 million tons a year and is expected to exceed 30 million tons by the year 2000. It will be over 40 million tons by the year 2005.

Propylene gas can be polymerized by heating it under a pressure of 15 atmospheres at temperatures from 50 to 90 Celsius in a solvent such as heptane, which is a liquid hydrocarbon with seven carbon atoms. This is called the slurry process. It can also be done without solvent, but then it requires pressures of 20 to 40 atmospheres, and this is called the bulk process. This method has the advantage of not having to reclaim solvent, but the disadvantage that it cannot be used to make so-called block copolymers.

It is also possible to polymerize propylene in the gas phase in a fluid-bed or stirrer-bed reactor at pressures of 8–35 atmospheres. The polymer is separated from the unreacted gas using a cyclone, which spins the stream of gas so that this can be recycled. The process is again more cost effective than the solvent process, and it can be used for making the full range of polymers.

All three processes can be operated continuously, and they all depend on Ziegler-Natta catalysts, named after the German chemist Karl Ziegler of the Max Planck Institute, and the Italian chemist Giulio Natta of the Polytechnic Institute of Milan. Ziegler discovered these catalysts in the 1940s and Natta developed them in the early 1950s. In 1963 they shared the Nobel Prize for chemistry for their work, which had transformed the polymerization of ethylene and propylene making the plastics they formed widely available.

The new generation of catalysts based on zirconium and titanium compounds was developed in the 1980s, and they are starting to replace the older catalysts. The new polymers they produce have a narrow range of chain lengths, with almost no wasteful short chains, and they also carry over less in the way of metal residues from the catalysts. The catalysts cost more, but the polypropylene they produce has an even better balance of impact strength and stiffness because it has a different *tacticity*. The new catalysts have transformed polypropylene because they can produce polymer chains of uniform length, with the methyl groups, which are attached to alternate carbon atoms of the chain, arranged in regimented arrays. These methyl groups give the polymer an extra dimension. If they all line up pointing in the same direction then the polymer is said to be *isotactic*. If alternate methyls point in opposite directions then it is *syndiotactic*. Strands of these polymers can mesh neatly with one another, like the teeth of a zip fastener, and the material is hard, tough and opaque—but not much use. On the other hand if some of the methyls are randomly orientated, i.e. the polymer is *atactic*, then we get a degree of irregularity which makes it much more flexible—and much more useful.

The early polymerization process produced up to 10% of the unwanted random polymer which had to be removed and was often dumped or burnt, but then was found to be a useful component of bitumen blends. The new polymer catalysts produce only 3% of the atactic type, but now a demand has grown for this such that by 1996 there was a world shortage of this former waste product!

We can increase the range of polymers even more by means of a little ethylene. Adding this produces a copolymer, of which there are two types called random and block. The result is polymers that can be longer and stronger, runny or sticky, rubbery or rigid. Around the world chemical companies are gearing themselves up to produce these new materials in the next few years, and we can expect to see them in some very unusual guises.

Random copolymer is made by adding ethylene gas to the propylene gas so that the two are polymerized together. The effect is to make the final material less crystalline, softer and more flexible, and to endow it with greater clarity. Random copolymer is used for disposable drinking glasses, cassette housings and bottles, and is formed by rapid cooling. Transparent film is made by stretching a sheet of random polypropylene copolymer after it has been extruded, and this is used for wrapping cigarette packets, food and clothes. It is also used to make adhesive tape.

Block copolymer, on the other hand, is made from partly polymerized propylene which is further extended by being joined together with sections of random polymer formed at a later stage. The result is a superb rubber which is tough and flexible, and remains so down to temperatures as low as −40 Celsius.

Chemically pure polypropylene has little going for it. It is lightweight, flammable, colourless, softens easily when warm, and it weathers badly. Indeed, tropical sunshine reduces unprotected polypropylene to a powder within a year! A dramatic improvement in properties comes by adding stabilizers. Some protect the polymer when it is molten, and some are added to shield it from the damaging ultraviolet rays of the Sun. In the careful hands of the polymer chemist polypropylene is made heat stable and sterilizable, impermeable yet flexible, transparent and perfectly safe. For example, by taking polypropylene sheeting just below its melting point of 170 Celsius and then moulding this by means of pressure, it is possible to make rigid food containers such as margarine tubs, whereas normal polypropylene would be considered too soft for such applications. Polypropylene meets the food packaging regulations of the European countries, as well as the US Food and Drugs Administration, and we come across it as yoghurt pots, sweet wrappers and snack packets. It is also widely employed for packaging medical products and appliances, such as vials and disposable syringes.

The ability of polypropylene to be stretched, and thereby improve its properties, accounts for some of its major uses as tapes and fibres. Tape polypropylene is the single biggest consumer and is woven into sacks, carpet backing, ropes and twines. Fibre polypropylene is made by extruding the molten polymer through fine nozzles, and it is used to make carpet pile, blankets, upholstery, wall covering, underwear and sports goods. There are several types of spun fibre, some of which are suitable as sewing thread, netting, filters, and even engineering constructions. Non-woven polypropylene fibre ends up as nappy (diaper) liners and tea-bags.

Because we come into physical contact with polypropylene we want it to be free of hazards, and it must meet the basic requirements of being unbreakable, non-toxic and non-flammable. Although bulk polypropylene is slow to ignite it will burn, but it can be made flame-resistant by adding flame retardants. If we want polypropylene to cope with rough handling and not to deform if it gets hot, we can add inorganic support materials such as chalk, mica and powdered glass. The plastic casing of the domestic equipment such as irons and kettles, for example, relies heavily on such modifiers.

Polypropylene is eminently recyclable. The automobile industry is well ahead in this respect, with polypropylene bumpers (fenders) a major source of material for recycling, along with battery cases. Other polypropylene artifacts that are recycled are milk and beer crates, chairs and textiles. Even when recycling is not feasible, polypropylene waste can be burnt as a high energy fuel in municipal incinerators to generate steam and electricity.

Portrait 5

Going to extremes on Earth and in the Heavens—Teflon

The previous two portraits are of plastics which perform as well as natural materials, but there are situations which demand far more than nature seems willing to give: the non-stick frying pan, the stain-repellant fabric and surface coatings that could survive in space.

On 20 July 1969, Neil Armstrong set foot on the Moon. When some people questioned the enormous cost of the trip, $14 billion, the US National Aeronautics and Space Agency (NASA) pointed to the more down-to-earth benefits it would bring. Thus was born the popular belief that the Moon-walk was a giant leap forward not only for mankind, but for the non-stick frying pan as well. This assertion is still widely believed, but it is a myth. In reality the Moon-landing would have been impossible without the coating of non-stick frying pans. If anything, the footsteps on the Moon were a giant leap forward for Teflon.

Teflon is one of the trade names for the polymer poly(tetrafluoroethylene), which is abbreviated to PTFE in the trade. This had first been made thirty years before the Moon visit: it was discovered by a 27-year-old chemist, Roy Plunkett, at the DuPont research laboratories at Deepwater, New Jersey. (Plunkett died in 1994, aged 83.) His polymer was destined to change the world, but perhaps not in any way he could possibly have conceived. Its first big role was in the production of the atomic bombs that fell on Hiroshima and Nagasaki in August 1945.

The Teflon story began on the morning of Wednesday 6 April 1938, when Plunkett opened a cylinder of the gas tetrafluoroethylene, which he was using to make CFCs. He was puzzled why a cylinder supposedly holding 1000 g of the gas only released 990 g. The explanation was to be found in 10 g of a curious white powder which he fished out with the help of a piece of wire. Plunkett realised that this was a new polymer, and investigations showed that it consisted of chains with about 100 000 carbon atoms, each with two fluorine atoms attached.

The new plastic had some remarkable properties: it was not attacked by hot corrosive acids; it did not dissolve in solvents; it could be taken down to −240 Celsius without becoming stiff and up to 250 Celsius without affecting its performance. Nor was that all: it could also be heated to over 500 Celsius without burning, and it had a peculiarly slippery feel. It was this which was to be the secret of its commercial success, and worldwide production is now about 50 000 tons per year with a value of £400 million ($600 million).

DuPont christened the new PTFE plastic Teflon, and this is the name most people know it by. The fluorine component of Teflon starts life as the mineral fluorspar, which is calcium fluoride (particularly fine specimens are sometimes called Blue John). Heated with sulfuric acid this yields hydrofluoric acid, which is reacted with chloroform, and when the product of this reaction is heated to 600 Celsius, it forms the tetrafluoroethylene gas from which PTFE is made.

The non-stick frying pan was a technological triumph, achieved by

Louis Hartmann back in the 1950s. He sought to bond PTFE to aluminium, and found a way to do it. The trick was to treat the metal surface with hydrochloric acid, apply PTFE as an emulsion, and then bake the pan at 400 Celsius for a few minutes. The acid etches tiny pits into the surface of the metal, and the PTFE flows into them. When the pan is heated the PTFE polymerizes into one continuous film of Teflon that is held tight to the surface by the polymer trapped in the millions of cavities on the surface.

The French company that invented the non-stick frying pan called itself Tefal, from the words Tetra Ethylene Fluorine Aluminium, and it went on to dominate the world's non-stick cookware market. Their first non-stick frying pans went on sale ten years before the trip to the Moon.

The year that saw the walk on the Moon also saw the launch of a remarkable fabric made of PTFE, and which is sold as Goretex. In 1969 Dr Bob Gore, of Maryland, found a way of expanding PTFE by heating and stretching the polymer to form a membrane. This created invisible pores in the film—billions per square inch—and these are small enough to keep water droplets out, but big enough to allow water molecules of sweat to escape. The Goretex film is widely used for wet weather gear and sportswear, where it is sandwiched between the outer fabric and the inner lining.

Goretex is ideal for golf suits, but many players of this game are middle-aged and they may have expanded PTFE inside their bodies as well. Artificial veins and arteries made of Goretex are a standard treatment for cardio-vascular disorders.

Teflon enters our everyday life in other ways: as fabric roofing for sports arenas; as stain-repellants for clothes, chair-covers and carpets; as plumber's tape for sealing joints in water pipes and central heating; on the underside of irons; and as dental floss. As you read this your fingers may be picking up PTFE from the page. PTFE scrap from industry is re-used by grinding it to a microfine powder and then adding it to printer's ink to make it flow more smoothly.

As we have seen, not all the uses of PTFE have been so innocent. Shortly after its discovery, it was in demand for the Manhattan Project, the aim of which was to make an atomic bomb. Its chemical inertness meant that it could stand up to fluorine gas, the most reactive of all molecules. Large quantities of this gas were needed to make uranium hexafluoride, from which the fissionable isotope uranium-235 can be separated. By 1942, and despite its cost, Teflon was being manufactured to make items which had to have the ability to resist the corrosive effects of raw fluorine gas, and today the chemicals industry relies on PTFE-coated tanks and vessels to hold highly corrosive chemicals.

Another project where money was of secondary importance was the race into space in the 1960s. The environments of extreme cold, low pressures, and the corroding effects of activated oxygen in the upper atmosphere required a material with unearthly properties, and PTFE was the only one suitable. Without it there could have been no voyage to the Moon.

Portrait 6

Getting rid of unwanted PETs—poly(ethylene terephthalate)

When the fabric called polyester first appeared in the 1950s it was seen as revolutionary because it was crease-resistant. Today we are more likely to encounter it as the bottle in which we buy a fizzy drink, and here it has almost supplanted glass as the preferred container, because it is lighter to carry, easier to handle, cheaper to transport, and safer to stack and use. The demand for this polymer has grown with our thirst for such beverages, and especially colas, which can now be sold in 'party' sizes—thanks to PET.

PET is short for poly(ethylene terephthalate), which is the older name given to the polymer whose correct chemical name is poly(ethene-1,4-benzoate). The increased demand for this plastic has come about because PET resembles glass, not only in its crystal clarity but also in its ability to provide an air-tight container over a long period, which is especially important for foods. Storage demands we keep oxygen out, because this will cause spoilage through oxidation, and for fizzy drinks we need to keep a pressure of carbon dioxide in, or the drink will go flat.

Most plastics have poor barrier properties when it comes to gases, but PET is an exception, hence its use as bottles, jars, and containers for foods as diverse as drinks, sauces, cooking oil, vinegar, honey, nuts, jams and wines. PET containers can also be used for cosmetics, toiletries and paints.

PET has a wider market than food containers: all photographic and X-ray films are made from it, as are video and audio tapes. A growing use for PET is in packaging for medical products such as ampoules and swabs. Here it scores over other materials because PET-enclosed products are easier to sterilize by irradiation.

PET was discovered in 1941 by two chemists, Rex Whinfield and James Dickson, who worked in the small research laboratory of the Calico Printers' Association in Manchester, England. They heated together ethylene glycol, better known as antifreeze, and dimethyl terephthalate at 200 Celsius and got a sticky mass of poly(ethylene terephthalate). They noticed that it gave long, strong fibres when drawn, and that the polymer threads were unaffected by boiling water—something they had not expected. Chemically, PET is described as an ester, and it depends upon the ester units to form the chains of polymer—and esters are normally quickly decomposed by water.

Nevertheless, their polymer was perfectly stable and it was launched after World War II as new fibre, which they named Terylene. Polyester was found to be particularly good in blends with natural fibres, especially cotton. One version of the new polymer fibre was called Crimplene, and this was invented by Mario Nava who lived in Macclesfield, England. It was manufactured by subjecting the yarn to a bulking process which resulted in a fabric that does not crease and is easy to launder. Crimplene is still very popular with regular long-distance travellers, who appreciate its uncrushability, although its popularity among older women, who appreciated that clothes made from it were easy to wash and needed no ironing, threatened its status as a fashion fabric for many years.

Even today most polyester production goes into textiles, and world output of the polymer now exceeds 1.8 million tons, of which the USA accounts for 700 000 tons, and Europe about 500 000 tons.

PET, the term commonly used for the polyester destined for packaging, is manufactured from pure terephthalic acid and monoethylene glycol, which react to form the starting material, bis-2-hydroxyethyl terephthalate. This is heated at around 200 Celsius under vacuum, with a catalyst, and as it melts it polymerizes to form PET resin. This is then subjected to further polymerization in the solid state which makes the polymer chains grow longer, and the whole becomes a crystalline clear plastic.

Turning this into bottles is a two-stage process. First, chips of plastic are injection-moulded to give a 'preform' which looks like a test tube. This is then reheated to just above its softening point before being blown into a mould to give the desired shape. This results in the polymer chains packing together more closely, with the result that the bottle is less permeable to gases. Billions of such bottles are now manufactured each year.

PET bottles are used for fizzy drinks (50% ends up this way), mineral waters (20%), edible oils (5%), fruit juices (5%) and others (10%). Yet good as they are, PET bottles were not welcomed everywhere. Despite its many benefits PET failed to meet the levels of perfection demanded by Germany's Greens, and in the 1980s they kept their country PET-free because they reasoned that glass was more environmentally friendly. Glass was seen as recyclable, which they regarded as of prime importance, despite the terrible accidents caused by broken bottles. When finally the introduction of PET bottles could be resisted no longer, there was imposed on each one of them a 50 pfennig deposit, reclaimable if the bottle was returned to re-use. Refillable PETs are now well established in Germany, the Netherlands, Austria and Scandinavia, and a typical PET bottle can be recycled up to 20 times.

These PET bottles are bigger and heavier than the one-trip bottles preferred in other countries, like the USA, where used PET bottles are collected but not refilled. They are melted down and turned into film suitable for other types of packaging, or into polyester fibre. In the USA over 30% of PET bottle resin is recycled into such products as carpets, duvets, anoraks, bristles for paint brushes and felt for tennis balls. Hulls for boats are being made of it, but the most eye-catching nautical use for old cola bottles has been the sails of tall ships. Europeans are currently less geared to recycling PET, but even so 400 million PET bottles are collected each year, and the total should rise as the public becomes more aware of the fascinating transformations that recycling brings. For example, five two-litre bottles can be turned into a tee-shirt, while a thousand will produce enough carpeting for a normal-sized living room.

Some PET is recycled chemically, in other words it is depolymerized into its starting materials by heating the plastic under pressure in methanol. The dimethyl terephthalate and ethylene glycol can then be purified, added to the feedstock, and returned for polymerization to make virgin PET. Care needs to be taken to ensure that there must be no PVC components to a PET bottle that is destined for recycling. These two plastics are completely incompatible—a little of one destroys the strength of the other.

PET meets other environmental criteria as well as recyclability. There is virtually no pollution emitted during its manufacture, and if it is incinerated it gives off a useful amount of heat. PET bottles have also got lighter in recent years. When they were first introduced in the 1970s a 1.5 litre cola bottle weighed 60 g; today the same sized container weighs only 44 g.

PET bottles are 25% more energy efficient than those made of other materials. For example, it takes 100 kg of oil to make a thousand one-litre PET bottles, but 250 kg of oil to make a thousand glass bottles. And it is not only during manufacture that energy is saved. A delivery truck can carry 60% more lemonade or cola and 80% less packaging if it is loaded with PET-bottled drinks rather than glass-bottled ones.

Whether beer drinkers can be persuaded to buy ales and lagers in plastic bottle remains to be seen. This market has been the toughest nut for plastic containers to crack, although in the UK the apple-based drink, cider, is mainly sold in PET bottles. But beer remains a challenge, because it is sensitive to oxidation and PET by itself cannot prevent some oxygen getting through. The answer is a bottle with outer and inner layers of PET, sandwiching a barrier layer, such as poly(ethylvinylidene alcohol), to provide an air-tight seal. This is 300 times less pervious to oxygen than PET, but it still has not displaced the aluminium or steel can of four-packs.

Portrait 7

Sexy and safe—polyurethane

The portraits of the two previous two plastics, Teflon and PET, showed us how these polymers could help us in ways that neither natural polymers nor glass could. The next polymer also deals with a few specifically human problems, such as family planning and energy-efficient fridges.

A factory in Cambridge, England, is turning out a new type of condom called Avanti, which sells for as much as £1 ($1.50) each. These are revolutionary because they are made from polyurethane, the plastic better known as lightweight insulation and upholstery padding. The polyurethane condoms are twice as strong as the traditional latex ones, so that they can be made much thinner; they are completely transparent, and slightly larger. Tests showed that 80% of users preferred them, reporting increased sensitivity. The new condoms are non-allergenic, unaffected by lubricants and provide an effective barrier against sperm and all sexually transmitted bacteria and viruses, including HIV.

Most people's idea of polyurethane is of the spongy material used for cushions and mattresses, or the lightweight rigid foam used in insulation panels. Our cars are full of polyurethane, and we save fuel because it saves weight. It provides comfort as the padding of seats, soundproofing under the carpets, and safety as cushioned dashboards and steering wheels. Polyurethane can also be a rubbery material suitable not only for condoms, but also for wellington boots, the soles of trainer shoes, and for Lycra swimwear and hosiery.

Chemists make polyurethane by reacting molecules having alcohol groups with those having isocyanate groups. As the molecules mix, a strong chemical bond quickly forms, binding them together and giving off heat. If a volatile liquid is also present this will form gas bubbles in the plastic, expanding it as it sets, like a sponge cake in an oven. Depending on the chemicals used, and the extent of bubbling, the final product can be a flexible foam, ideal for furniture, or a rigid foam, suitable for fridges and wall insulation. Polyurethane foams weigh so little because they can be 95% gas.

Polyurethane foam is helping to solve South Africa's housing needs. For those who still have to live in shacks of corrugated iron and plywood, a temporary answer is to spray the building with polyurethane, which makes them livable in by keeping out insects and the heat of the Sun, and making them soundproof. The polyurethane cladding is then painted with a fire-retardant resin, which also protects the plastic from the Sun's damaging ultraviolet rays. A house can be treated for as little as £120 ($180). Nor will the investment be wasted when people are rehoused: they can cut the polyurethane into panels with a knife and use it as insulation in their new home.

Forty per cent of polyurethane output goes into rigid polyurethane, and 30% into flexible foam. In the past people have been alarmed by the flammability of polyurethane itself, and by the use of CFCs as the foaming agent needed to make it. Today there is little to worry about on either count. Polyurethane meets all the new fire safety standards, which have become increasingly stringent, and none is made using CFCs, at least in Europe and the USA. (CFCs are still being used for polyurethane foams in developing countries such as China, which, rather strangely, were exempted from the Montreal Protocol when it was agreed to phase out these chemicals in the West.) CFCs have been replaced by ozone-friendly materials such as hydrofluorocarbons (HFCs), pentane or carbon dioxide.

The international chemical company ICI has recently developed a flexible foam in which the bubbles are made using water, which reacts with a little of the isocyanate used to make polyurethane, to generate carbon dioxide. They have also designed a new generation of insulation based on polyurethane foam, suitable for use under vacuum. This 'super insulation' is up to three times more effective than ordinary polyurethane, so it can be thinner—so fridges can be built smaller or have larger interiors.

However, it is the non-foam polyurethane which is providing the new markets for some of the five million tons produced each year. Lycra, the stretch fabric, has expanded into sports clothing and fashion in addition to its more conventional use in swimwear. For the first time, in the World Cup of 1994, footballs were coated with polyurethane making play faster, or so it was claimed. Polyurethane can be used as an adhesive to bind together other materials, and in this way old tyres are shredded and bonded to form athletic tracks and surfaces for children's playgrounds.

Nor, when its useful life is over, need the polymer be wasted. It can generate energy by being burnt as fuel in municipal incinerators, and has the calorific content of coal. Better still, it can be separated and recycled by reducing it to its constituent chemicals, which can then be reconstituted as a new batch of polyurethane.

Portrait 8

Film set fun and fast food flaw—polystyrene

Like polyurethane, polystyrene was once best known as lightweight insulation. It has also been berated because it was used to make throwaway packs for hamburgers. It is still used for disposable containers for hot coffee.

Environmentalists rightly campaigned against the throwaway culture, and they not only accused polystyrene users of increasing the mountain of waste, but they also accused polystyrene of being a threat to the ozone layer, on account of the CFC gases which were used to puff out the polystyrene polymer. Yet far from being a pollution threat, this plastic has now been redeemed as the most environmentally friendly of polymers, because every ton of expanded polystyrene manufactured for insulation will save three tons of heating fuel a year. Only 0.2% of oil goes to making expanded polystyrene, which is tiny compared to the 35% we burn to keep us warm. But expanded polystyrene is more than just about saving a family money on their heating bills—it may occasionally save their lives when it is used as safety padding in vehicles and crash helmets.

Polystyrene was discovered as long ago as 1839, but only went into commercial production in 1930. It is made from styrene monomer, which itself is made from benzene and ethylene. This is a colourless, oily liquid which boils at 145 Celsius, and it can be polymerized by heating a suspension of it in water, using peroxides to initiate the polymerization. The product is small beads of polystyrene which are then graded into various sizes ranging from 0.2 to 3 mm.

The beads soften at 94 Celsius and melt at 227 Celsius, giving this plastic a useful working range in which it can be softened with steam and shaped or moulded. The product has good electrical resistance properties and resists acids and alkalis. It is not affected by hydrocarbon solvents and alcohols, but it is soluble in many other organic solvents. Polystyrene has the disadvantage of burning easily, giving off clouds of smoke and melting, but this can be controlled by fire-retarding additives.

Polystyrene is a molecule with benzene rings bonded to every other carbon atom of the polymer chain. The benzenes explain the unusual features of polystyrene. They make it glass-like because benzene rings on one chain tend to attract those on another, making the plastic less flexible and more brittle than other polymers. However, this ensures a tighter packing of the polymer chains, and this results in a transparent material with a high refractive index, giving it the attractive sparkle of glass. The benzene rings also absorb ultraviolet radiation, which may be an advantage indoors when used to screen fluorescent lights which gave off ultraviolet rays. Outdoors the benzene rings are a disadvantage causing the polymer to become yellow and degraded due to the intensity of the ultraviolet radiation in sunlight. Then, like humans, it needs to be screened from those destructive rays.

Although expanded polystyrene foam is perhaps its most familiar form, there are actually three guises in which this remarkable polymer reaches the customer: as polystyrene, as expanded polystyrene, and as high impact polystyrene. Each has many uses, and the table shows how widely these have entered our lives.

Polystyrene by itself is used for a wide range of consumer products such as yogurt pots, fridge interiors, eye-shadow compacts and toys. Disposable drinking glasses for parties and picnics are crystal-clear, and sparkle like glass. Like glass they are brittle, but they do not fragment into dangerously sharp slivers. Slightly thicker polystyrene is used to make tape cassettes and CD cases. A film of polystyrene is used for the windows on the front of business envelopes.

High impact polystyrene, as its name implies, is tough. This is achieved by adding up to 10% of polybutadiene or of styrene-butadiene copolymer. The resulting material is no longer transparent, but it is much stronger and can be used to make packaging for food. High impact polystyrene can be extruded into sheets and then softened and moulded into refrigerator door liners, dinnerware, dairy food containers, camper/ trailer covers and circuit-breaker housing. While ordinary polystyrene has poor resistance to fats and oils, high impact polystyrene is fine and this is the preferred packaging for spreads and margarine.

The versatility of polystyrene

Products made from polystyrene itself
Brushes, combs, razors
Cosmetic compacts
Disposable drinking glasses
Scientific equipment (disposable pipettes), etc.
Video, audio and CD cassettes

Products made from expanded polystyrene
Building materials, insulation for buildings
Containers for hot drinks
Boxes for transporting fresh fish packed in ice
Lining for crash helmets
Protective packaging for electrical goods

Products made from high impact polystyrene
Combs and coat-hangers
Circuit breaker housing
Food trays, margarine tubs, yogurt pots
Fridge interiors
Toys

Expanded polystyrene offers a package of benefits that makes it unique. We come into contact with expanded polystyrene when we unpack domestic equipment which is protected by this remarkable lightweight plastic. Because it is so light, expanded polystyrene packaging reduces transport costs. We may hardly give it a second glance, little appreciating that it has been craftily shaped to fit the product and to give it maximum protection during transit. (Sadly, all we can then do with it is throw it away.) For the same reasons expanded polystyrene makes an ideal protective layer inside crash helmets.

Safety is also the reason why set designers choose expanded polystyrene for dramatic and violent scenes. We wince when we see actors on screen crushed by chunks of falling rock or masonry, and we hold our breath as the hero struggles to pick up unbelievably heavy objects to rescue a trapped child. All good clean expanded polystyrene fun, but this same stuff also has some more serious uses as a building material. It is water-resistant and very buoyant, so it finds use in marinas, lifebuoys and pontoons. Civil engineers use expanded polystyrene granules as a filler in the concrete for embankments, bridges, elevated motorways, dams, docks and harbour walls, without in any way compromising the physical strength of these structures.

However, it is the insulating ability of expanded polystyrene which has the biggest impact on our lives. We are intrigued, when we hold a piece of expanded polystyrene, to discover how warm it feels to the touch, and we appreciate that our home can be made more energy-efficient with it. Good insulation not only keeps homes warmer in winter, but also keeps air-conditioned rooms cooler in summer; and fridges and freezers work more efficiently all year round thanks to a lining of expanded polystyrene.

Thin-walled cups and food containers designed to hold hot drinks and meals account for relatively little expanded polystyrene, but it was these which became the target of environmental campaigns in the 1980s. Not only was such packaging immediately discarded, often as litter in the street, but the gases used to expand or 'blow' the polystyrene were the ozone-damaging CFCs. Today the blowing agent is pentane, a volatile hydrocarbon which poses no threat to the ozone layer.

When Dr Martin Hocking of the University of Victoria, Canada, did a full cradle-to-grave analysis of expanded polystyrene drinking cups, he came up with a rather unexpected result: those made out of expanded polystyrene turned out to be more environmentally friendly than those made of paper. According to Hock's eco-audit, more chemicals were needed to produce a cardboard cup than a polystyrene one. Making paper cups also needs more water, and more steam, and over ten times the electricity.

Not only does expanded polystyrene appear to offer environmental advantages, but it is now also losing its throw-away image. Polystyrene is recyclable, and the expanded polystyrene industry has long been recycling its own waste. So-called transit packaging, a major use of expanded polystyrene, is now recycled across Europe. Several electrical appliance manufacturers collect used packaging and it is being turned into items such as cassette casings and plant pots. Some is recycled to the construction industry, where granulated expanded polystyrene is used for drainage or in the manufacture of lightweight building materials such as bricks, concrete and plaster.

The expanded polystyrene which can't be recycled may still provide a final bonus: its energy. If it ends up in a municipal incinerator then like other plastics it releases as much heat, weight-for-weight, as coal or oil. If this energy is converted to electricity or hot water for a local heating scheme then expanded polystyrene completes its brief life in a highly environmentally friendly finale.

Portrait 9

Stronger than steel—Kevlar

The reason why TWA flight 800 crashed in the sea near Long Island in July 1996, killing 229 passengers and crew, may never be known for certain. At the time, and just as the world was awaiting the start of the Atlanta Olympic Games, it was assumed that the people who lost their lives were victims of a bomb in the aircraft's hold. This is what had happened to Pan Am flight 103, which exploded over Lockerbie, Scotland, in 1988 killing 270 people. But an aircraft might survive a explosion in its baggage hold if this were sealed with Kevlar panels.

Kevlar is the plastic that can stop a bullet, which is why it is used in bullet-proof vests. It already protects planes by lining the engine compartment to limit the damage that might be caused should a turbine blade fly off, and because of its strength and lightness it has been used in the framework of the Boeing 757. Whether a Kevlar baggage hold could contain the blast of a bomb remains to be seen. The UK Defence Evaluation Research Agency (DERA) has worked on a £5 million project to find out whether such protection is feasible, and they have carried out a full-scale trial on a pressurized aircraft, at DERA's explosions facilities at Fort Halstead. These tests showed that the Kevlar panels would deform without breaking, while containing the bomb fragments. About three tons of Kevlar would be required to protect a plane, and this would add around £35,000 a year to fuel costs.

Kevlar was discovered in 1965 by Stephanie Kwolek, working for the US chemical giant, Du Pont, on a project to design a fibre that had the heat resistance of asbestos and the stiffness of glass fibre. However, there were problems because of the nature of the polymer and the special solvent needed to make it, which was discovered to be carcinogenic. These delayed its launch until 1982, by which time it had cost $500 million to develop. It was hailed as 'a miracle in search of a market'—and it is still searching for the elusive mass market, despite its unique properties and the original hope that Kevlar would replace the rayon fibre and steel wire reinforcement in tyres. Du Pont refuses to reveal how much is produced at its plants in the USA, Japan and Maydown in Northern Ireland.

The polymer consists of long strands of benzene rings interconnected with amide groups, very like those in protein. Kevlar forms when a benzene with two amine groups reacts with another benzene with two acid chloride groups. What gives the polymer its remarkable strength is its regularity of structure. In most fibres the strands of polymer are a random, tangled mass, but in Kevlar the attractive forces between strands are so strong that they line up in parallel rows, making flat sheets which pack as rigid layers on top of one another.

This regularity creates problems because it makes the polymer insoluble, although it will dissolve in pure sulfuric acid, from which it can be extracted unharmed. This is one way in which Kevlar can be processed. In addition to being almost immune to chemical attack, Kevlar is also fire resistant, flexible and lightweight. When it is spun into fibres and heat treated, the polymers get even stronger, and they are used for military armour, space suits, safety gloves and fishing rods. Kevlar is incorporated into tennis racquets, skis and running shoes. The plastic is five times as strong as steel and more elastic than carbon fibre, and it has pushed the performance limits of such equipment well beyond that of traditional materials.

Another sporting area where Kevlar's remarkable strength has been employed is in Formula One racing cars, although less is now being used because of its limitations. According to Brian O'Rourke, chief structural engineer of Williams Grand Prix Engineering, the benefits of Kevlar's high tensile strength are offset by its poor compression strength, and it is difficult to paint. Even so, because it has good rigidity-to-weight ratio it is incorporated into the laminate used to reinforce the driver's survival cell, where it provides excellent protection in case of a crash.

Few plastics have the package of benefits which Kevlar provides. When it fails, then it does so progressively rather than catastrophically, thereby providing another margin of safety. Unlike many plastics, it does not become brittle at low temperatures, even as low as −70 Celsius, and optical fibres are coated with it if they are exposed to the severity of mountain conditions. Nor is Kevlar affected by long exposure to weathering or to the sea, and three years immersed in either boiling water or in hydrocarbon solvent left it unchanged. Kevlar is flame resistant, self-extinguishing and gives off little smoke, so it is preferred for conveyor belts, especially in mines, and for hoses used in the chemical industry and engines. Mooring ropes for tankers are made of Kevler rather than steel, but perhaps the most dramatic use of Kevlar has been in body armour, flak jackets and head gear, which are not only lighter than other forms of protection but can also be tailored to fit.

It seems a little odd that a polymer with all these assets has failed to find more of a role in modern life. That does not mean that some day we will not be thankful for its unique combination of properties.

◆

The portraits of Gallery 5 have all been members of the same family, the carbon-based polymers, and like all families the differences are often more apparent than the resemblances. And while in their early years some of these have been rather undisciplined, they have all eventually grown up to be useful members of society.

GALLERY 6: LANDSCAPE ROOM

ENVIRONMENTAL CONS, CONCERNS AND COMMENTS

An exhibition of molecules that stalk the world

A HUNDRED YEARS ago, if you talked about protecting the environment you meant preventing floods or forest fires. Homes and farms could be ruined and families wiped out by a flash flood, a surge tide, or a raging fire. Meanwhile in industrial regions the skies were polluted with fumes, smoke and smog, rivers were little more than open drains and slag was piled up in great heaps. People complained but there was little they could do, because their livelihoods depended on the very industries which were causing the pollution. Excesses were curbed, but change was painfully slow.

Fifty years ago, when you spoke of protecting the environment you meant controlling urban sprawl and cleaning up the wastes of industry. The climate of opinion now favours quicker changes, and much has been achieved since then: slag heaps have been sculpted into grassy knolls, derelict sites have been demolished and turned into sport centres or superstores, rivers now support fish and wildlife abounds on their banks. The belching smoke and choking fogs of coal-burning industries are only memories. And while the air in cities is now fouled by traffic fumes, there are signs that this pollution too will disappear as cars become cleaner.

People today have other environmental concerns. They want action taken on different kinds of pollution. It is not enough to pull down old factories, gas works and foundries and to turf over the site: we want the soil beneath to be decontaminated too, so that homes can be built there and children can play safely in gardens. People want power to be generated without causing acid rain. They want all rivers and lakes to be so clean that people can fish from them or swim in them.

When it comes to breathing, we have little choice. The air we breathe comes with the neighbourhoods in which we live and work. Clearly, we have some control: we can avoid traffic fumes, and change the ventilation of the rooms we are in, but even so the mixture that we are taking in is still a cocktail of gases, some of which are not natural, and some of which may be hurting us. Little wonder then that we are easily alarmed by stories of dangerous gases in the air.

When it comes to other parts of our environment, we do have some control. We may worry about the water we drink, and if we think that the water that comes from our kitchen tap is not for us, then we can go to the supermarket and buy any of a dozen brands of bottled water. There is a lot of other action that we can take at an individual level, and so we find people willing to reduce household waste by reusing things before discarding them, and to reduce municipal waste by recycling as much paper, plastic and metal as possible. Nor do they want such waste to be dumped in holes in the ground and covered over: they prefer it to be burnt, so it can generate electricity and hot water for local homes. They also want houses that are better insulated, home appliances that use less energy, family cars that need less fuel and last longer, or—better still—they want cheap public transport and dedicated cycle lanes.

Despite all this effort in the right direction, for some it is still not enough. There are idealists who are fired by the vision of a green paradise, a return to a Garden of Eden where all is natural, self-sustaining, and in harmony with nature. They dream of a world of small towns and villages separated by natural woodland and wild open spaces. With the help of science all this can be achieved, and without forgoing the benefits of a secure food supply, comfortable homes, a good health service, a fulfilling educational system, rewarding jobs, and a rich cultural life. I am sure that it will come one day, but it will only come as a result of the efforts of chemists, biochemists and biotechnologists.

Meanwhile, we have to live in the world as it is, with its ever-growing population and ever-larger cities. In this Gallery I have brought together a few portraits that touch on environmental concerns. The first of these are about the gases in the air we breathe. Then we will examine a couple of pictures of molecules that seemed to provide a better environment, but are accused of making things worse.

Portrait 1

The air that we breathe—oxygen

The most important gas in the atmosphere is oxygen, which makes up around 21% of the volume of dry air. Without enough oxygen to breathe we die, and this might happen if we are in a confined space and the oxygen gets depleted, or if the air pressure is too low because we are very high up. There the air may still contain 21% oxygen, but the pressure might be too low for our lungs to extract it. But even at the top of the highest mountain, where the air is thin, there is still enough oxygen. Early explorers thought it would be otherwise, but they were later shown to be wrong.

On 29 May 1953 Tenzing Norgay and Edmund Hillary became the first men to climb Mount Everest, which they did with the help of oxygen cylinders. Forty years later Harry Taylor, a 33 year-old ex-SAS officer, climbed to the summit alone, without extra oxygen. In 1975 the first woman to scale the peak, Junko Takei of Japan, took an oxygen cylinder.In May 1996 the late Alison Hargreaves became the first woman to achieve this feat without the aid of oxygen.

We need oxygen so our body can generate energy, and we draw upon the ready supply of this gas in the atmosphere. However, there is a lower and an upper limit to the amount of oxygen in the air if it is to be considered safe. If we are not to suffocate, the oxygen level must stay above 17%; if we are not to burst into flames it must stay below 25%.

We can breathe oxygen-enriched air, as many sick people do, but if we are surrounded by it we are in danger. Hospital patients inside oxygen tents have suffered horrific burn injuries when they have tried to light a cigarette. The three astronauts destined for the first manned Apollo flight in Earth orbit were burned alive in minutes in their spacecraft on 27 January 1967, when fire started in the oxygen-enriched air of the cabin. In October 1969 at South Shields, in north-east England, the same fate befell a group of ship repairers in the hold of the *Lady Delia*. They were using a drill normally worked by compressed air, but which had inadvertently been connected to a supply of oxygen. When the critical 25% was exceeded, and one man lit a cigarette, it burst into flames which spread to his overalls. As his workmates rushed to help him they too ignited. Within minutes four men lay dead, and seven were badly burned. This mysterious case of multiple spontaneous human combustion was solved by Professor Ian Fells of the nearby University of Newcastle-upon-Tyne, who investigated the apparent mystery and eventually discovered the wrongly connected hose.

But it is too little oxygen which is generally the threat to life, and which brought the Biosphere project in Arizona to a premature end in January 1993. Eight people had been sealed into the glass-walled ecosystem in December 1991, to see if it was possible for humans to sustain life in a space station or on the Moon. Within a few weeks they were gasping for breath as the oxygen in the air fell below 17%. Somehow 30 tons of it had disappeared, and it was thought that it had probably reacted with iron in the soil.

Oxygen is attracted to the iron or haemoglobin in our blood, and thereby efficiently transported to where it is needed. (Most, but not all, species use iron as the oxygen carrier. Spiders and lobsters use copper, which is why their blood is blue.) Thanks to haemoglobin, a litre of blood can dissolve 200 ccs of oxygen, fifty times as much as will dissolve in the same volume of water. But if the amount of oxygen in the air decreases then so does the amount in the blood, and even though our heart may be pumping as quickly as it can to make up for the deficiency, it cannot sustain this extra output of energy for long, and we die.

Molecules of oxygen gas consist of two oxygen atoms, but the bond between the atoms still puzzles chemists. It appears to be a double bond, and yet the molecule has two rogue electrons which means that it is a so-called 'free radical'.

Oxygen gas will liquefy at −183 Celsius and the liquid is magnetic, as Michael Faraday discovered in 1848 when he spilled some and watched it run towards the poles of a magnet; it behaves like this because of these two free electrons. In theory these should make it react instantly with anything it touches, and yet we know that oxygen is a relatively unreactive molecule, otherwise it could not have built up over millions of years until it comprises a fifth of the Earth's atmosphere. Even when it enters our body it does not immediately react chemically with its target molecules, but needs an enzyme catalyst to make it react.

There are a *million* billion tons of oxygen gas encircling the globe, and all of it is produced as a by-product of photosynthesis in plants. The seven billion tons of fossil fuel we burn each year consumes around 24 billion tons of oxygen, which is only 0.00024% of the total, and plants replace most of it. Even if plants did not replenish the oxygen in the atmosphere, it would take over 2000 years at the present rate of depletion for the oxygen level to fall from 21% to 20%.

Our brain must have oxygen to function, and without it this vital organ will begin to die within minutes. Less well known is that too much oxygen will poison it. This threat is not appreciated by many sports divers, according to Kenneth Donald of Edinburgh University, Scotland, who has made a life-long study of the subject. In his book *Oxygen and the*

Diver Donald warns against breathing pure oxygen below 25 feet, since this can lead to convulsions and several divers have drowned. Instead of using compressed air, amateur divers such as undersea photographers, bounty hunters and archaeologists have taken to using so-called nitrox mixtures, which is air with a boosted oxygen content—but it too can be dangerous. Nitrox is a mixture of nitrogen and oxygen, and was developed by the British Navy in the World War II for divers disposing of mines, because it allowed more time underwater while avoiding oxygen poisoning and the decompression sickness, the 'bends'. Today professional divers breathe a costly mixture of oxygen and helium, which enables them to work safely down to 500 metres and below.

Oxygen is produced industrially by distilling liquefied air, and is either made on site, or delivered via a pipeline, or transported in specially insulated tankers. In the USA production is 25 million tons a year, and in the UK it exceeds 4 million tons. Over half goes to making steel, about a quarter to making ethylene oxide which is turned into antifreeze or polyester for bottles and fabrics (see Gallery 5), and the rest is used as the gas itself, in medical care, or to purify sewage and so prevent environmental disasters like the one in Paris in 1992. A violent storm caused raw sewage to flood into the river Seine, and this rapidly used up the oxygen in the water and killed all the fish. There are now giant pumps to bubble 15 tons of oxygen gas a day into the Seine.

Who first discovered oxygen? The credit usually goes to Joseph Priestley, who was born in Leeds, England. He was a nonconformist preacher, a left-wing intellectual who supported the aims of the French Revolution, and an amateur chemist, who specialized in studying gases. He discovered oxygen in 1774 after he moved to Lord Shelburne's estate at Calne in Wiltshire, and it was there that he heated mercury oxide and collected the gas which it gave off. He breathed his new gas and reported how light-headed it made him feel. He also noted that a mouse could survive much longer in this new gas than in ordinary air. Priestley moved to Birmingham but there his house and laboratory were ransacked by a right-wing mob. Perhaps not surprisingly, he eventually emigrated to the USA.

Little did Priestley realize that Carl Scheele at Uppsala, Sweden, had made oxygen a few months earlier, but had failed to gain the credit that was his due because the publisher to whom he sent his manuscript did nothing to publish it. Neither Priestley nor Scheele was responsible for naming the new gas. 'Oxygen' was chosen by the great French chemist Antoine Lavoisier, and it means 'acid-forming'. Lavoisier thought, wrongly, that this element was an essential component of all acids.

But could there have been a previous discoverer of oxygen? There is evidence that oxygen was produced 150 years earlier. How else do we explain a remarkable event that occurred in London in 1624, when King James and his subjects turned out in their thousands to watch a new wonder of the age: a submarine. This remarkable craft consisted of a wooden framework covered by a watertight, greased leather skin. It was manned by 12 rowers whose oars protruded through sealed ports. With its Dutch inventor, Cornelius Drebbel, on board with a few other passengers, it sailed for two hours underwater from Westminster to Greenwich, a distance of several miles. (The Admiralty were not impressed and advised against its adoption.) This mysterious journey was still being talked about 40 years later by no less a scientist than Robert Boyle, of Boyle's Law fame. He wrote that one of the passengers, then still alive, had said that when the air in the submarine had been consumed, Drebbel was able to refresh it with purer air from a container. It has been suggested that this purer air must have been oxygen.

One explanation is given by Zbigniew Szydlo in his book *Water Which Does Not Wet Hands*, in which he says that Drebbel was conversant with the work of the Polish alchemist Michael Sendivogius, who lived from 1566–1636, and who knew of a gas which he referred to as 'the aerial food of life.' 'Water Which Does Not Wet Hands' was Sendivogius' code name for nitre. Sendivogius had observed that when nitre (the old name for potassium nitrate) was heated, gases were evolved. Gentle heating of this salt produces oxygen. In those days nitre was collected from the walls of cellars and latrines, where it grew as white crystals, or from the leachings of manure and soil. Nitre was gathered on a commercial scale because it was needed to make gunpowder.

Nitre's curious ability to produce oxygen might have been known to John Mayow (1641–1679), an Oxford chemist and early fellow of the Royal Society of London. He wrote about 'nitro-aerial particles' which came from nitre when it was heated, and this phrase too is thought to refer to oxygen. It has even been suggested that the alchemists' Elixir of Life was not a liquid as popularly supposed, but might have been this secret gas, oxygen.

Portrait 2

So much and so unreactive—nitrogen

Although oxygen is the essential gas of the air, the most abundant is nitrogen. In its own way nitrogen is just as important as oxygen, because nitrogen is also an element essential for life. Unlike oxygen, however, which can be highly reactive given the right conditions for combustion, nitrogen shows little inclination to react. Despite this inertness, it has to be made to react because nitrogen is needed for every strand of DNA, every fibre of muscle, and every enzyme in every cell of every living thing on this planet.

Nitrogen reacts with oxygen when a bolt of lightning passes through the air, and thunderstorms can bring a lot of nitrogen to earth each year as nitrate, which is the form in which plant roots absorb it from the water in the soil. This nitrate is nowhere near the amount needed to support all the vegetation there is. Certain plants such as beans, and marine organisms such as algae, carry enzymes called nitrogenases that can induce nitrogen from the air to react and so 'fix' it as ammonia gas. Until chemists devised a method of fixing nitrogen as ammonia in the early part of this century, these enyzmes supplied almost all the nitrogen for the planet's biomass. Humans can now fix nitrogen by reacting it with hydrogen to form ammonia, which is also an excellent fertilizer, but this chemical reaction requires high temperatures and pressures.

How do the nitrogenases fix nitrogen so effortlessly? For 50 years chemists, biologists and biochemists struggled to find out, and in the 1990s they laid bare these enzymes' inner secrets.

Atmospheric nitrogen molecules consist of two nitrogen atoms joined together by a very strong chemical bond—a triple bond, in fact. To turn nitrogen into ammonia the two atoms have to be separated and each bonded to three hydrogen atoms. For this to happen the enzyme has to provide six hydrogens and six electrons. In fact it provides eight hydrogens and eight electrons, and makes not only two ammonia molecules but also a molecule of hydrogen gas.

What puzzled early investigators was that nitrogenase itself consists of two large proteins, and both are essential for the process. The smaller protein contains four iron and four sulfur atoms; the larger protein has about twelve iron atoms, twelve sulfurs and two atoms of the rare metal molybdenum. Douglas Rees and researchers at the California Institute of Technology at Pasadena have performed the difficult trick of separating the smaller protein, growing a crystal of it and working out its exact chemical structure by bombarding it with X-rays. This revealed a ring-like protein with four irons and four sulfurs clustered together rather like a gemstone protruding from the ring. The purpose of this protein is to provide electrons and pass them to the larger protein—the 'gemstone' touches a special site on the larger protein.

Another US group led by Jim Bolin at Purdue University concentrated its efforts on the larger protein, which picks up the nitrogen gas molecule. Their research revealed the structure of this half of the enzyme, and showed the arrangements of its iron and sulfur atoms, but in addition it revealed the positions of the all important molybdenum atoms, which attracts the nitrogen from the air.

Research also provided an answer to the puzzle of the hydrogen gas given off during nitrogen fixation: the hydrogens are present in the site attached to the molybdenum, and these are displaced when a nitrogen molecule comes along. The larger protein of nitrogenase waits till a nitrogen molecule wanders into its cavity where it displaces a hydrogen whose function is probably to protect the site from chemical attack by other molecules. Once inside the site the nitrogen sticks to the molybdenum atom and bonds a hydrogen atom to itself, at the same time absorbing an electron from the surrounding iron atoms. The larger protein then pumps in more hydrogens and the smaller protein adds yet more electrons to the molybdenum site until the nitrogen has produced one ammonia molecule which then leaves the cavity. The remaining nitrogen is given the same treatment until it has the three hydrogens needed to convert it to ammonia, and off it goes as well. The protein then protects the active site with fresh hydrogen until the next nitrogen comes along.

All this sounds rather complex, but now the secret is out chemists might be able to model similar molecules that will catalyse the reaction of nitrogen and hydrogen at ordinary temperatures, thereby saving the vast amount of energy which currently goes into nitrogen fixation. Alternatively, biotechnologists and genetic engineers may find a way to transfer the gene which encodes for the nitrogenase enzyme to other plants, so that they become self-fertilizing, at least as far as nitrogen is concerned. Either way, the object will be to ensure that human food and crop production is limited to as small an area of this planet as possible. This way we can leave more land untouched and available as natural habitat for the wildlife that shares this planet with us.

Portrait 3

The lazy loner that does a lot—argon

The discovery of argon is attributed to Lord Rayleigh and William Ramsay, who announced it in 1894 and then released no further details until the following year. By so doing they were eligible to enter a competition organized by the Smithsonian Institution in Washington, DC, for 'some new . . . discovery about atmospheric air.' They won the $10,000 prize, worth about $150,000 today. Lucky argon! But it was second time lucky.

Argon was first discovered unwittingly in 1785 by Henry Cavendish of Clapham, South London. (The same Henry Cavendish is also mentioned in portrait 8 in this Gallery.) He was interested in the chemistry of the atmosphere, and passed an electric spark through a mixture of air and oxygen, absorbing the gases which formed. Yet no matter how long he sparked the air there remained 1% of its volume that would not combine chemically. What he did not realize was that he had stumbled on a new gaseous element. For over a century his observations were not understood—but they were not forgotten.

What prompted the second discovery of argon was the mysterious behaviour of nitrogen: why did the density of this gas depend on where it came from? The nitrogen extracted from the air had a density of 1.257 g per litre, whereas that obtained by decomposing ammonia gas had a density of 1.251 g. Rayleigh and Ramsay knew that either atmospheric nitrogen must contain a heavier gas, or chemically derived nitrogen contain a lighter gas. The latter explanation was most unlikely, so they concentrated their attentions of the nitrogen derived from air.

Ramsay passed a sample of supposedly pure nitrogen from the air over heated magnesium, with which it reacts to form a solid, magnesium nitride. Like Cavendish, he was left with about 1% of the volume which would not react. This gas was 30% denser than nitrogen. When they examined its atomic spectrum they observed new lines which could only be explained by a new element. Ramsay and Rayleigh based their name for the unreactive gas on the Greek *argos*, meaning idle—and argon it became. Ramsay won the Nobel Prize for chemistry in 1904.

Argon, which constitutes 1% of the air, is now an important industrial gas. Hundreds of chemical plants around the world extract it from liquid air. A typical plant processes 375 tons of air a day, and is controlled by computer and run by only a handful of technical staff. It separates the air into oxygen, nitrogen and argon, which are shipped out as liquids in tankers holding 20 tons per load.

Argon is particularly important for the metals industry. The steel industry uses argon as an inert gas to stir molten iron while oxygen is bubbled through it to adjust the carbon content. Argon is also used when air must be excluded to prevent oxidation of hot metals, such as with molten aluminium. If this metal is to be welded it needs to be protected against the oxygen of the atmosphere. This can be done by using an electric arc welder which uses direct current to create a spark which melts the welding rod. This is surrounded by a flow of argon and a typical welding apparatus needs around 10–20 litres of argon gas a minute. Atomic energy scientists protect fuel elements with it during refining and reprocessing. The alloys for high grade tools require ultrafine metal powders, and these are produced by directing a jet of liquid argon, at −190 Celsius, at a jet of the molten metal. Some smelters prevent toxic metal dusts escaping to the environment by venting them through an argon plasma torch. In this, argon atoms are electrically charged to reach temperatures of 10 000 Celsius and the dust particles are turned into a blob of molten scrap.

Argon lasers are used by surgeons to weld arteries and kill tumours. Their intense blue beam is also used by chemists to probe molecular states which exist for only a trillionth of a second.

Some consumer products contain argon. This is the gas in the gap between the panes of sealed double glazing, where it improves the insulation because it is a poorer conductor of heat than ordinary air. Argon is also the gas inside fluorescent tubes and light bulbs; in the latter it dissipates the heat of the incandescent filament while not reacting with it. Illuminated signs glow blue if they contain argon, and bright blue if there is a little added mercury vapour. The most exotic use of argon is in the tyres of luxury cars. Not only does it protect the rubber from attack by oxygen, but it ensures less tyre noise when the car is moving at speed.

Many of these uses of argon rely on its chemical inertness—nothing will induce it to react with any other material, no matter how high the temperature to which it is heated, nor how strong the electrical charge passed through it. So far it has resisted all attempts to make it bond to other atoms: argon gas consists entirely of single argon atoms. Even compounds which contain argon, the so-called argon clathrates, hold it only as atoms trapped in holes in the lattice of a larger molecule.

There are several trillion tons of argon swirling around the globe, where it has slowly built up over billions of years. Most of it has come from the element potassium, which has a radioactive isotope, potassium-40, with a half-life of 1.28 billion years. Only 117 atoms of potassium in a million are potassium-40, and when a particular atom's time is up its nucleus can either emit a β-ray and turn into calcium-40, or it can capture one of its own electrons and turn into argon-40. Only one atom in ten prefers the latter choice, but given the great age of the Earth, around 4.6 billion years, there has been time enough for all the argon we now have to be formed. It also explains why the atomic weight of argon (element 19) is greater than that of potassium (element 20), because most argon is argon-40 (with a relative atomic weight of 40) whereas most potassium is potassium-39 (with a relative atomic weight of 39).

If the radioactive potassium-40 is dissolved in the sea, dispersed in the soil or a component of a living organism, then the argon escapes into the atmosphere. But if the potassium is trapped in the rocks or the Earth, it stays put. By measuring the ratio of potassium to argon in a mineral, it is possible to date it.

Argon is one of a group of elements called the noble gases. Most of these were discovered by Ramsay and Morris Travers during the years 1895 to 1898. Three were extracted from the air: neon, krypton and xenon, which they named from the Greek words *neos* (new), *krypton* (hidden) and *xenos* (stranger). A fourth, helium, was again a rediscovery.

This gas, and the lightest of the noble gases, had been detected 30 years earlier by Pierre Janssen when he went to India to study a total eclipse. He recorded a yellow line in the Sun's spectrum which he could not explain, but which indicated an unknown element. The astronomer Sir Norman Lockyer named it helium from the Greek *helios* (Sun), and speculated that it could not exist on Earth. Ramsay found it in 1895, by extracting it not from air, although it is more abundant than krypton and xenon, but from a uranium ore which gave off bubbles of helium as it dissolved in acid. (When a radioactive atom emits an α-particle it is emitting the nucleus of a helium atom. This quickly picks up two electrons to give helium gas.)

Portrait 4

Too high and too low for comfort—ozone

Experience teaches us that not all friends, families, cars, gourmet meals and holidays are good, nor that all next-door neighbours, in-laws, fast food and airports are bad. And so it is with molecules in the environment. Some are targeted as bad, but even these may have their good points. Ozone in the lower atmosphere is polluting, while ozone in the stratosphere is good because it protects us from the Sun's damaging rays. It is the former ozone that this molecular portrait highlights. We may view it as the picture of a villain, but our grandparents and their grandparents saw it as a hero.

Chemicals can have their seasons, just like fashions. What one age admires as fine, another will reject as folly, and a good example of this is ozone. Each year at the end of winter the ozone in the upper atmosphere becomes depleted and we rightly worry because this gas shields life on

Earth from the dangerous ultraviolet radiation of the Sun. Each summer down here at the surface we worry as the ozone in the air we breathe increases, because this gas is damaging to living things.

A century ago ozone was also something to worry about, and for exactly the opposite reason: it was thought there was not enough of it around. Ozone was deemed to be natural, wholesome and invigorating, and the very locations where its levels were highest proved this: up in the mountains and along the coasts. There, the air was fresh and clean—there was the place to go to convalesce and take your holidays, away from the smoky cities which were supposed to be lacking in ozone.

William S. Gilbert, who was later to achieve fame with Arthur Sullivan for a series of Victorian musical comedies such as *The Mikado* and *HMS Pinafore*, wrote the poem 'Ozone' in 1865. This neatly encapsulated the popular belief of the time about the gas. Its third and last verses go as follows:

But if on Ben Nevis's top you stop,
You will find of this gas there's a crop—but drop
To the regions below,
And experiments show
Not a trace of this useful ozone is known,
Not a trace of this useful ozone!

It's because I'm an ignorant chap, mayhap,
And I dare say I merit a slap or a rap,
But it's never, you see,
Where it's wanted to be,
So I call it Policeman Ozone—it's known
By my friends a Policeman Ozone!

It is difficult to imagine an age when chemicals in the atmosphere were regarded as benign and something you could poke fun at. Gilbert thought ozone was abundant high in mountains but depleted down in the cities of the plain. Today the reverse applies: we are threatened by it in cities and worry about its lack on high.

Such was the esteem in which the Victorians held ozone that they had generators pumping it into churches, hospitals, theatres and even their underground railways. Today we are a little wiser, and know that ozone is a lung irritant. There is a natural low level of ozone in the air, about 20 parts per billion (ppb), which in summer can increase to 100 ppb as a result of sunlight acting on nitrogen dioxide from car exhausts.

In the summer of 1976 a value of 260 ppb was recorded in the UK, the highest ever, and well over the legal limit for the occupational exposure, which is 100 ppb. The ozone damages the macrophage cells in our lungs so they are less capable of consuming and destroying bacteria, and the irritant effects of ozone also make it harder to breathe.

Ozone is a blue gas with its own instantly recognizable 'metallic' smell. It can be condensed into a blue liquid or frozen into a violet-black solid, but this is rarely done because both are dangerously explosive. Ozone consists of three oxygen atoms joined together in a V-shaped molecule; it is not a stable form of oxygen and quickly reverts back to ordinary oxygen with its two atoms. This is especially rapid in the presence of a catalyst like charcoal.

Ozone comes from the Greek *ozein*, meaning smell, and there are classical references to the sharp odour of ozone during thunderstorms. We can often smell ozone near high voltage electrical equipment and electric sparks. It is a very strong oxidizing agent, which is why it is very good at destroying microbes.

The industrial uses of ozone are increasing. It is used in the chemicals industry to make plasticizers for polymers like PVC, and by pharmaceutical firms to make sterile water. It is used to preserve food, and to kill bacteria in bottled mineral waters. Ozone for such processes is made on site and used immediately.

There are two ways of making ozone industrially, and these mimic the natural ones. The usual method is to pass air through concentric glass tubes with metallized surfaces across which is applied a discharge of 15 kilovolts and 50 Hertz. The gas emerging from such treatment contains 2% ozone. Another method, used when only low concentrations of ozone are needed for sterilizing or disinfecting, is to expose air to ultraviolet light.

In the next century ozone will become the preferred chemical for sterilizing water for drinking and swimming pools. Ozone is a more potent disinfectant than the commonly used chlorine gas: it not only destroys common pathogens, but it will also kill crypto-sporidium, the recently discovered microbe that causes dysentery and can survive mild chlorination. Sterilizing with ozone is more complicated than with chlorine, where merely bubbling the gas through the water to give the necessary concentration will suffice. When water in a swimming pool is disinfected with ozone, the process is in several stages. First, a saturated solution of ozone in water is prepared, and this strong solution is then used to disinfect water from the pool. Before the water is returned to the pool, excess ozone is removed by passing it through a charcoal filter. Finally a little chlorine is added, and the water is put back in the pool. The chlorine keeps the water sterile, and unlike ozone, it remains active for a long time. The level of chlorine in public pools has to be high to counteract the public's curious lapses of personal hygiene, even though the chlorine can affect the breathing of some people, and make their eyes sore.

Ozone is ideal for sterilizing the cooling water in power stations, where the growth of bacteria interferes with heat transfer and so reduces efficiency. In France ozonization has been the preferred method of water treatment for domestic supplies for 80 years, and there are over 600 ozone units in operation.

Ozone is also being used to clean up waste water from sewage, especially if it is to be discharged near holiday beaches, and units are in operation not only to kill bacteria but also to kill the sulfurous smell of methyl mercaptan (whose portrait is to be found in Gallery 1) which comes from volatile sulfides released by bacteria. These sulfides are oxidised by ozone to odourless sulfate.

Happily none of the above uses adds to the ozone pollution of the lower atmosphere. Researches into the effects of ozone on plants and crops show that the ozone pollution in the lower atmosphere is damaging plants, especially in rural areas. Ozone weakens plants and makes then more susceptible to other natural stresses such as insects, fungi and frost. Research in the USA has revealed that ozone levels of 50 ppb can cut yields of crops such as wheat by a sixth, at an estimated cost to US agriculture of $3 billion annually. It is not difficult to imagine that a lot more food is lost when levels reach 250 ppb, as they sometimes do.

Curiously ozone levels in cities, where traffic pollution is generated, is quite low. But this is still the cause of high ozone levels over the surrounding countryside and farm land, where a combination of nitrogen oxides, hydrocarbons and strong summer sunlight conspire chemically to generate ozone.

As William S. Gilbert pointed out, ozone is never where we want it to be, although today we want just the opposite to what he wanted, which was more down here rather than up there. The ozone hole which appears over the Antarctic each spring is worrying because each year it seems to get larger, and while it is a perfectly natural phenomenon, its expansion is thought to be caused by gases we have put into the atmosphere. Ozone is depleted by chlorine atoms, and any gas which contains chlorine and can survive in the atmosphere for a long time will eventually diffuse up to the ozone layer and there release its chlorine atoms under the intense ultraviolet radiation from the Sun. Each chlorine atom then has the capacity to destroy a million ozone molecules. The chlorofluorocarbon gases, better known as the CFCs, qualify as the chief ozone depleters and these were used for nearly 30 years in aerosols and fridges, and manufactured on the million-tone scale. Severe curbs have been introduced and the levels in the atmosphere are now falling, but it will be many years yet before the threat to the ozone layer has passed.

Portrait 5

Acid rain, vintage wine and white potatoes—sulfur dioxide

Breathing ozone may be bad, but breathing sulfur dioxide is worse. In December 1952 sulfur dioxide killed 4000 Londoners in the worst-ever 'pea-souper' fog which choked the city for five days, eventually spreading to cover an area of a thousand square miles. In those days all homes were heated by coal fires, which is where the sulfur dioxide came from.

Sulfur dioxide is generated wherever coal is burnt in industry or power plants, and to a lesser extent it is produced in cities, where other fuels are burnt. The whole planet has to live with the sulfur dioxide that drifts out of industrial areas, and spews from active volcanoes. It oxidizes in the air, dissolves in the droplets of water from clouds, and then falls to Earth as acid rain.

Over 300 million tons of sulfur dioxide are released to the Earth's atmosphere every year. About half is of volcanic origin, and half comes from fossil fuels. It could be worse, because sulfur dioxide gas is produced when any sulfur containing substance burns. Thankfully the sulfur in natural gas, and to a large extent that in oil, is removed before burning, but it is less easy to remove sulfur from coal unless the coal is first gasified by treating it with steam and then burnt. A few plants capable of doing this have been built, but the more general approach is to burn the coal and then remove the sulfur dioxide from the flue gases.

Many industrial countries are now committed to reducing sulfur dioxide emissions by 70% of their 1980 levels within 25 years. Some are doing remarkably well, such as the UK which is well past the half-way mark. The success was achieved not so much by design, but as a by-product of industrial changes: most countries are moving steadily towards their target by phasing out old coal-burning power stations, or removing sulfur dioxide from flue gases with sprays of water and neutralising the acid with lime, but this is an expensive answer to the problem. The USA is trying to reduce sulfur dioxide by using a different approach—making the polluter pay. The aim of the US Environmental Protection Agency is to halve emissions to under 10 million tons per year by 2000. In April 1993 the EPA held the first auction at which it sold the rights to emit sulfur dioxide, most of which were bought by coal-burning power stations at an average price of $150 per ton of sulfur dioxide emitted.

Despite the worries about acid rain, and its effects on lakes and vegetation in areas near industries which burn brown, high-sulfur coal, it is generally quite beneficial to forests and fields. If that comes as a surprise, let me explain further. In 1992, the Finnish Forest Research Institute in Helsinki published a paper entitled 'Biomass and carbon budget of European forests' in the journal *Science* (volume 256, page 70, 1992). It reported the findings of Pekka Kauppi, Kari Mielikäinen and Kullervo Kuusela, who said that, contrary to what was generally believed, the trees of Europe are not dying in their millions, killed by acid rain, but are flourishing as never before—responding to the increasing amounts of carbon dioxide and sulfur dioxide in the atmosphere. The trees are the first signs that we are witnessing the long-awaited greening of the planet forecast by some scientists. Human activity adds 25 billion tons of carbon dioxide annually to the atmosphere, most of which can be accounted for, but six billion tons disappear without trace. The Finnish scientists believe that the missing carbon dioxide is going into trees, and that Europe's woods and forests account for about a tenth of it. If trees all over the world are likewise growing bigger, all the missing mass could be explained. What Europe's trees are doing is putting down larger and deeper roots, and to do this they are using the sulfate of acid rain as fertilizer. Acid rain may supply as much as 4 g per square metre per year of this nutrient, which is enough to stimulate growth in the poor soils on which many forests grow.

Another good side of sulfur dioxide is that it preserves our food, and it is regarded as safe because this molecule is a natural, albeit fleeting, part of our body's metabolism. Sulfur dioxide kills bacteria, is a good antioxidant, and prevents food from going brown. It is used on a large scale as a preservative, both as itself, or as sodium, potassium and calcium sulfite salts which form sulfur dioxide in solution. Sulfur dioxide is not generally permitted in meats because it destroys vitamin B_1, but there are exceptions such as sausages and the Scottish delicacy, haggis. Sulfur dioxide is widely used with fruits and vegetables because it preserves their natural colour—for example, peeled potatoes stay white. The manufacturers of those rather tasty vegetarian alternatives to hamburgers, the so-called vegeburgers, have been given approval by the UK's Ministry of Agriculture, Fisheries and Food to use sulfur dioxide as a preservative.

From the days of the Roman Empire, sulfiting, as it is known, has been a common method of preserving wine. The sulfur dioxide was generated by the simple act of burning natural sulfur; this was done near vats so that the grape juice in them could absorb the fumes. As little as 100 parts per million of sulfur dioxide is enough to prevent undesirable yeasts from multiplying while allowing desirable yeasts to flourish. When grapes are crushed the juice will spontaneously ferment due to the action of wild yeasts on the surface of the fruit. The sulfur dioxide suppresses these, and a cultured yeast is added to do the fermenting. These yeasts can survive high levels of sulfur dioxide—indeed, some actually produce sulfur dioxide themselves. Home brewers and winemakers use sulfur dioxide in the form of sodium metabisulfite tablets, which are also known as campden tablets. Wines may also be treated with more sulfite just before bottling, to prevent further fermentation, and this might require levels as high as 350 mg per bottle. Most of this sulfur dioxide eventually reacts with other components in the wine and disappears, but some young white wines have noticeable amounts.

The world's leading research unit looking into the effects of sulfur dioxide on food is headed by Dr Bronek Wedzicha of University of Leeds, England. Wedzicha's book, *Chemistry of Sulfur Dioxide in Foods*, claims that sulfur dioxide is the most versatile food additive available, and one of the safest: it controls every form of food spoilage—microbial decay, oxidative breakdown, and browning. It even extends the life of vitamin C. According to Wedzicha there are many ways in which sulfur dioxide can react with components of our food, but tests on rats have shown these to be safe.

Sulfur dioxide gas is easily recognized by its choking smell which irritates the nose and lungs, and it can seriously affect some people who are very sensitive to it. They may even experience the same reaction from soft drinks or young wines. Beers, lagers and ciders also have sulfur dioxide in them but generally much less, and for most drinkers its presence goes unnoticed.

Nevertheless sulfur dioxide is an irritant, and can affect those prone to asthma. The gas is sometimes used to test for this condition since it triggers a rapid response in mast cells in the air passages. These are cells in loose connective tissue surrounding blood vessels, and are to be found especially in the lungs and gut. Mast cells have granules containing histamines which they release in response to sulfur dioxide and other irritants, and this triggers the attack. In the USA, where there is also widespread use of sulfur dioxide, some deaths were even alleged to have been caused by it.

Despite the dire warnings of those who oppose food additives, sulfur dioxide can be counted as safe, because it is a natural chemical formed within our bodies during the metabolism of amino acids. Not only that, but we have an in-built safety system which can mop up any surplus sulfur dioxide and turn it into harmless sulfate. The capacity of this detoxifying process is such that it can cope easily with any sulfur dioxide that we consume.

Portrait 6

Too much of a good toxin—DDT

DDT was a devil in disguise. In the midst of World War II it appeared like a beacon of hope, and Winston Churchill spoke of this new chemical in a radio broadcast, describing it as 'an excellent powder . . . which yields astonishing results, and which will be used on a great scale by the British forces in Burma.' In 1944 the Allies had used DDT to stop an outbreak of typhus in the newly captured city of Naples.

DDT was eventually to save an estimated 50 million lives. In a war-torn world this was good news indeed, but until then the molecule had been a military secret, code-named G4. G4 sounded like a new wonder drug, but it was merely the insecticide DDT, whose initials stand for dichloro-diphenyl-trichloroethane. The molecule had first been made as long ago as 1874, by a chemistry student Othmer Zeidler. He had taken the chemical chloral, which used to be known for its rapid action as a sleeping potion (called knock-out drops or Mickey Finns), and mixed it with chlorobenzene in sulfuric acid. The result was a white precipitate of DDT crystals. Zeidler reported his new molecule, and that was that. Nobody noticed its remarkable insecticidal activity.

DDT was rediscovered in 1939 at the Geigy company in Switzerland by Paul Herman Möller, who was searching for new insecticides. He tested the powder and was amazed at how effective it was at killing all kinds of insects at very low doses. It was soon in commercial production, and in the next 30 years over 3 million tons of DDT were manufactured. Möller received the 1948 Nobel Prize in Medicine and Physiology for his contribution to human health.

DDT kills an insect by interfering with its nerve cells. The molecule unlocks a channel through the cell membrane which allows sodium atoms to flow in unchecked, causing the nerve to trigger repeatedly until the insect dies of exhaustion (animal nerve cells are not affected in this way). This 'excellent powder' was destined to save millions of lives by eradicating disease-bearing insects such as lice, which cause typhus, fleas which cause plague, and mosquitoes which cause malaria and yellow fever. DDT also destroyed crop pests such as the Colorado potato beetle, and it was much safer than the insecticides then in use which were based on the poisonous elements arsenic, lead and mercury. Yet today many people regard DDT as an equally dangerous toxin.

However, in the 1950s its achievements were impressive. A campaign to eradicate malaria from the then British colony, Ceylon (now Sri Lanka), began in 1948, when there were 2.5 million cases of the disease annually. Every home on the island was sprayed regularly with DDT, and by 1962 there were only 31 reported cases—the scourge of this ancient disease had apparently been lifted.

Not everyone approved of chemical insecticides, and that same year Rachael Carson's book *Silent Spring* was published. This passionately moving book was to become the bible of environmentalism. Carson referred to DDT as the 'elixir of death.' Within a few years people were lobbying for it to be banned, claiming it was killing wildlife, especially birds, causing cancer in humans, and building up in the environment because it was not biodegradable. Moreover, analytical chemists were able to detect this pesticide in tiny amounts and they had revealed that it was everywhere: in soil, in water, in our food, and even in human tissue.

In addition to these worrying claims, there was an even stronger scientific reason for phasing out this insecticide: the appearance of DDT-resistant strains of insects. These insects produce an enzyme that detoxifies DDT by removing a chlorine atom from the molecule. Today there are almost 500 species resistant to DDT—mute testimony to its over-use. It is still employed as an insecticide in some tropical countries like India, although restricted to 10 000 tons per year. The USA banned DDT in 1972, as did many other developed countries. The spraying of homes in Ceylon had ceased in 1964, but within five years there were again 2.5 million cases of malaria on the island.

In her book *Toxic Terror*, Elizabeth Whelan, president of the American Council on Science and Health, debates the pros and cons of DDT and questions the wisdom of banning this cheap and effective insecticide. She points out that DDT may have saved more lives than any other chemical. She also challenges many of the misconceptions about it, and says that there is no evidence from human studies that it causes cancer.

The original report that DDT was persistent in the environment was based on applying it to a plot of soil at ten times the normal level, and then keeping the soil dry and in the dark. The DDT did not degrade. However, in normal soil DDT is digested by microbes and its activity persists for only about two weeks. The microbes also deactivate it by removing a chlorine atom, and the same thing happens in sea water, where 90% of DDT disappears within a month. Nevertheless, DDT did accumulate in humans, and at the time when DDT was banned the average person had about 7 parts per million in their body. This came from their diet, because most food in the 1960s contained about 0.2 ppm of the insecticide. The DDT concentrated in fatty tissue and was excreted only slowly; the half-life of DDT in the body is 16 weeks.

Such levels of DDT were never dangerous to our health. The World Health Organization's guidelines for a safe level of DDT intake is 255 mg per year—about ten times the amount which consumers were exposed to when DDT use was at a maximum in the late 1960s. We know from accidents and suicide attempts that people can drink a glassful of insecticide fluid, containing about 4000 mg (4 g) of DDT, without harmful effects. The fatal dose for a human is thought to be about 30 g of the pure material.

Curiously, some insect species can tolerate even higher doses of DDT. One is the Brazilian bee *Eufriesia purpurata*, which is found in the Amazon. These bees actively seek out DDT and collect it. Indeed, some of them have been found to have over 4% of their weight as DDT, which would be the equivalent of a human having six pounds of the insecticide in their body. To these bees it appears to be a sex attractant, which is perhaps not too surprising because some sex pheromones are very similar to the DDT molecule.

So what should we do about DDT? Clearly there is little that can be done to reinstate it, but there are lessons to be learned from its story. Perhaps we could still allow it to be used, on a limited scale, against a few insects that have not become immune to it, but it will never again be used as freely and unthinkingly as it was in the 1940s and 50s.

Portrait 7

Mad cows and madder chemists—dichloromethane

Bovine spongiform encephalopathy (BSE) is commonly called mad cow disease. It appeared in the UK in the mid-1980s and almost destroyed the beef industry, which was an important part of the economy. What was perhaps most frightening was the way in which the disease had been able to jump from one species of animal to another: from sheep to cows and cows to antelopes (that happened in London Zoo). It affects cats and finally humans, where it appears as a form of Creutzfeldt–Jakob Disease (CJD) in young people. What is rather sad is that none of this need ever have happened, had not a simple solvent been wrongly suspected of being environmentally dangerous.

BSE, like the disease scrapie, is a wasting disease of the brain. Scrapie appeared in sheep more than two hundred years ago, and may well have been around even longer than that. When offal from slaughtered sheep, including those with scrapie, was processed into cattle fodder and fed to cows they became infected with a new disease. The disease-carrying agent is not one of the usual pathogens such as a bacterium, virus or fungus: a small protein called a prion is believed to be the cause, and this explains why it is unlike other diseases, and why it can cross the species barrier.

The horror of disease jumping from one species to another, and thereby finding a totally unprotected population in which it can run riot, haunts the human mind and threatens the human population. It may have happened in the past, and would explain how virulent epidemic diseases can suddenly appear and wipe out much of the human race. The Black Death is the most infamous of such catastrophic diseases. In his apocalyptic and anti-science novel, *Ape and Essence*, which is set in the year 2108AD, Aldous Huxley presents a vision of a future after the nuclear holocaust of World War III. The human race is blighted by genetic mutations and terrible diseases caused by radiation:

> Glanders, my friends, Glanders—a disease of horses, not common among humans. But, never fear, Science can easily make it universal . . . .

Huxley goes on to describe its awful symptoms. Glanders is a contagious disease which can jump the species barrier, from infected horses to humans who touch them, but it is not common. Yet Huxley was playing on a real fear, and one which has proved justified with BSE. But the BSE tragedy might never have happened, had a useful chemical not been wrongly condemned as unsafe, and its planned use in the processing of abattoir waste not been abandoned. The chemical in question is a common DIY solvent which is still widely available and used in paint brush restorers and paint strippers.

Chemists in Britain had discovered that the solvent, dichloromethane (DCM), was ideal for extracting fat from grieves, the dried form of abattoir waste which is converted to high-protein cattle fodder. Grieves is animal offal that has been pulped and heated under pressure at 120

Celsius to drive off the water it contains, after which it needs to have its fat removed. This can be done by treating it with a solvent, and the ones previously used were either hexane, a dangerously flammable solvent, or trichlorethylene, which was safer but was found to contaminate the product by reacting chemically with part of the protein.

A pilot plant which used DCM as the solvent had already been built, and it was producing high grade fat and cattle cake, both free of the BSE agent. Plans were made to introduce the new process in place of the old methods, but before this could be done a report from the US Environmental Protection Agency reported that DCM caused cancer in mice. Faced with this alarming discovery, the British firms that processed abattoir waste abandoned the new solvent, and went over to a non-solvent process instead. This used lower temperatures of around 80 Celsius to render the grieves, and then these were pressed to extract the fat. Unfortunately the BSE agent survived the new treatment, and cattle began to eat the infected fodder.

The solvent DCM was also attacked from another quarter. Environmentalists accused it of damaging the Earth's atmosphere because, like CFCs, it contains ozone-depleting chlorine atoms.

Research has since shown that DCM does not cause cancer in humans. Nor does it damage the ozone layer, because it is quickly oxidized to form products that are easily washed out of the air by rain. Meanwhile, throughout these scares DCM continued to be the active ingredient in DIY paint brush restorers and paint strippers. It has a remarkable ability to penetrate the hardened surface of paint films and lift them off.

DCM is also known in industry by its older name of methylene chloride. It is a clear, volatile, non-flammable, colourless liquid with a pleasant odour. It is a simple molecule with two hydrogen atoms and two chlorine atoms attached to a carbon atom. It is used industrially on a large scale to clean metal surfaces and to dissolve oils, fats, waxes, resins, rubber and tar. It is essential for the manufacture of viscose yarns, cigarette filters and cellophane, which are made from DCM solutions of cellulose acetate.

ICI's Chlor-Chemicals plant at Runcorn, Cheshire, is Britain's largest producer of DCM, and they make it from methanol. Production worldwide is around a million tons a year, with ICI producing a fifth of this. DCM was first introduced as a safer alternative to ether—an equally volatile, but dangerously flammable liquid that was common in hospitals and laboratories until the 1960s. It was tried as an anaesthetic, but it was not widely used. It has, however, proved very popular in other ways, and a high-purity grade of DCM is used extensively by pharmaceutical and cosmetic manufacturers.

As with all volatile solvents, DCM is tightly regulated. The safe working level in air is 100 ppm, well below the 2000 ppm level that causes headaches and vomiting, and the 20 000 ppm that will cause death. Most DCM that enters the body is expelled on the breath, but some is converted to carbon monoxide and this could affect people with a heart condition. Splashes of DCM on the skin can sting alarmingly, but the effect soon wears off if the affected area is bathed with water, and there is no permanent damage.

What undermined the use of DCM for making cattle fodder was the announcement that it caused the development of cancers in a particular strain of mice who were exposed to high levels of vapour. What was not generally realized was that these mice are specially bred for test purposes to be sensitive to cancer-forming chemicals, as we saw in Gallery 1. Research on rats and hamsters showed no increased risks of cancer, and epidemiological studies on 6000 people who had worked with the solvent over many years showed no increased susceptibility either.

What really clinched the argument that DCM was harmless to humans was the work of Trevor Green at Zeneca's Central Toxicology Laboratory at Macclesfield, England. He had researched DCM for ten years, and had discovered the reason for the special sensitivity of the particular strain of mice that were affected by it. They have high levels of an enzyme, glutathione-s-transferase, in the nucleus of each cell which can activate the DCM to form a metabolite. This metabolite mutates the cell's DNA and triggers off the cancer. Although rats, hamsters and humans also have this enzyme in their bodies, it is not located in the nucleus of the cell and so does not act as a carcinogen.

There are no natural supplies of DCM, apart from small amounts given off by erupting volcanoes, and the current atmospheric level of 0.05 parts per billion can be attributed almost entirely to human activity. Even if more is manufactured, this level is unlikely to rise because DCM is destroyed by light and oxygen, and has a life-span of only nine months in air. It is no threat to the ozone layer, nor does it cause photochemical smog over cities, and government scientists have concluded that it has little effect as a greenhouse gas.

The earlier conviction of DCM as a dangerous pollutant now looks to have been a sad miscarriage of justice. Indeed, had it not been wrongly convicted it might have prevented BSE, and so saved the lives not only of millions of cows but also of several humans.

Portrait 8

Water, water, everywhere—H_2O

Water has fascinated scientists down the ages. The ancient Greek philosopher Thales thought that water was an element, and it was regarded as such until 1774, when Henry Cavendish, whom we encountered in Portrait 3 earlier in this exhibition, showed that it was a compound of hydrogen with oxygen. Since then it has become one of the most investigated of all chemicals, but it is still one of the most puzzling.

Fewer things could be simpler than a water molecule, H_2O, consisting as it does of two hydrogen atoms attached to an oxygen in a V-shaped arrangement. Yet nothing is as complex as water in its behaviour. For example, H_2O should be a gas like its sister molecule H_2S, hydrogen sulfide. Moreover, when it freezes at 0 Celsius its solid form, ice, floats instead of sinking. Water expands when its temperature falls below 4 Celsius, and expands most of all as it turns into ice. (In Gallery 8 there is a portrait of antimony, another material that expands as it solidifies.) This expansion explains why pipes burst when they freeze in winter, and why ice cubes chink so comfortingly in most drinks rather than sitting sullenly on the bottom—although this is where they reside if the drink is mainly alcohol.

We should be grateful that ice floats, because if it did not there would be almost no life on this planet. If the waters in which life started had frozen solid in winter, life would have been snuffed out straight away. Because water freezes from the top down the ice actually protects the creatures which live underneath it. Only a few microbes that can exist in ice would have survived.

The reason why water is a liquid lies with its two hydrogens, which act as a kind of chemical adhesive, sticking one molecule to another through hydrogen bonds. In the liquid state of water these bonds make and break continually, but as ice forms they lock into an open framework, like a great honeycomb of molecular cells. This framework is lighter than water, and so it floats. If it froze into a tightly packed solid the world would be a very different place, and the North Pole would be a solid chunk at the bottom of a new ocean.

Recently, chemists have been able to make water do strange things by turning it into supercritical water, which they do by heating it well above its boiling point. Although water boils at 100 Celsius, this is only strictly true at sea level. At the top of Mount Everest it boils at about 75 Celsius because of the reduced air pressure. At the bottom of the deepest mine it boils at a few degrees above 100 Celsius. If we continue to increase the pressure, we can increase the boiling point up to a maximum of 374 Celsius at a pressure of 220 times atmospheric pressure. Above this critical temperature liquid water cannot exist, no matter how high the pressure. It becomes a so-called 'supercritical' fluid, in which it is a gas but with the properties of a liquid.

As such it will dissolve almost anything, even oils, and when it does the volume of fluid can suddenly shrink to a half or less. This happens because supercritical water tends to pack tightly around other molecules. More strangely still, organic materials will 'burn' in it, in other words will be destroyed into simpler molecules. Treatment with supercritical water has been suggested as an alternative to incineration for disposing of sewage sludge, which dissolves to form a crystal clear, odourless, germ-free solution.

When oxygen gas is pumped into supercritical water it becomes a powerful oxidizing agent, able to break down some of the most persistent toxic wastes. US researchers at the Los Alamos National Laboratory in New Mexico are developing this as a way of disposing of unwanted rocket fuels, explosives and chemical weapons. Chemicals react fast in supercritical water, with some reactions going 100 times faster than under ordinary conditions. The trouble with supercritical water is that it is capable of slowly corroding almost any metal, even gold, and the problem faced by researchers is to find a material for pressure vessels that will resist it.

There are other ways of boosting water's activity than heating it under pressure until it goes supercritical. Ultrasound, whose frequency is too high for humans to hear, does remarkable things to water, creating tiny bubbles, inside of which there exist extremely high temperatures and pressures for a fraction of a second when the bubbles collapse. Under such conditions a water molecule inside the bubble will cleave one of its hydrogen atoms to form the highly reactive hydroxyl radical. This will then react with any other molecule it meets, and in this way dangerous or intractable materials in the water can be got rid of. Sonochemistry, as it is called, can even eliminate CFCs, which are difficult to dispose of because they were designed to be non-flammable and chemically unreactive. This was why they were widely used for 40 years in aerosols, insulation foams, and cooling units. A group of Japanese chemists, headed by Yoshio Nagata, at Osaka University, Japan, have demonstrated that the CFCs collected for disposal from old fridges and air-conditioning units can be converted to simple chemicals like carbon dioxide and hydrochloric acid simply by blasting them with sound waves in water at 20 Celsius.

Portrait 9

Water white and crystal clear—aluminium sulfate

When we draw a glass of water from a tap we want it to look crystal clear, yet the natural source from which a public water supply comes may be murky. While much of this cloudiness will settle out if the water stands in a reservoir, the customer wants all of it to be removed. To do this the water has to be treated with a flocculating agent, which causes even the finest particles to clump together so that they can be filtered out.

For over a century, water engineers have been using aluminium sulfate as a flocculating agent. When water is cloudy with silt and bacteria it can be made sparklingly clean by adding a small amount of slaked lime (calcium hydroxide) and aluminium sulfate. This combination precipitates solid aluminium hydroxide, which carries impurities down with it as it sinks. The effect of flocculation, and the insolubility of aluminium hydroxide, is such as to leave behind only 0.05 ppm of dissolved aluminium in the water—well below the 0.2 ppm suggested for drinking water. Surprisingly, flocculation can remove excess aluminium which may be present naturally. That the amount of aluminium left behind in the water is tiny is reassuring because there have been fears in the past that aluminium in the body can exacerbate the brain condition known as Alzheimer's disease. These fears were ill-founded as it turns out, but that is another story. When this portrait was painted it was believed that aluminium was highly dangerous.

Our picture is painted against a landscape of a small community in Cornwall, England, where in 1988 the residents of the village of Camelford were suddenly drawing tap water which contained very high levels of dissolved aluminium. On 6 July that year a driver arrived at the local water works, with a tanker of 20 tons of a concentrated acidic aluminium sulfate solution. Mistakenly he pumped it not into a storage tank, but straight into the mains. Within hours Camelford residents were complaining to the water company, but it had already discovered the mistake, and had immediately flushed the mains with fresh water. This discharged the aluminium sulfate solution into the nearby river Camel, where 50 000 fish quickly died.

Aluminium is not welcome in our drinking water, and certainly not at the level of 600 ppm that the residents of Camelford drew from their taps. But could aluminium seriously affect our health?

Aluminium is the most abundant metal element in the Earth's crust, and is found in combination with oxygen and silicon in rocks such as granite, and in soils, especially clays. Although it requires a lot of energy to extract the silvery metal from its main ore, bauxite, the effort is justified. Once extracted it can be used again and again with only a little extra input of energy, and today a large percentage of aluminium is recycled.

We probably use aluminium in more ways than any other metal. Aircraft, ships, containers, beer kegs, cars and cables are made from it. In the home it is found as window frames, cooking foil, pans and drinks cans. Aluminium is light and strong, and it resists corrosion because there is an aluminium oxide film on its surface which is tough and impenetrable. We may wear impure aluminium oxide in jewellery, as rubies, sapphires and topaz. The oxide itself is white, but metal impurities give it colour. We consume aluminium salts in our food, or in large doses as aluminium hydroxide indigestion tablets.

Aluminium sulfate is produced on a million-ton scale for use in the paper and water industries. It is a stable white powder made from aluminium hydroxide and sulfuric acid, but it dissolves readily in water, and a litre will hold over 350 g (13 ounces). Indeed, it is so soluble that it is mostly transported as a concentrated solution, which is how it arrived on that fateful day in Camelford.

For centuries people have used the mixed potassium aluminium sulfate salt, known as alum or potash alum, in many ways: as a mordant in dyeing, as a tanning agent, a cement hardener, a food additive, and even as an astringent to stop bleeding. No one suspected that aluminium might be a health hazard. Then in the 1970s doctors discovered that aluminium could present a serious health problem, when dialysis dementia was diagnosed. Patients on kidney dialysis machines suffered progressive brain damage and were dying. The cause was eventually traced to high levels of aluminium, which came partly from the large volumes of water required for dialysis, and partly from aluminium fittings in the machines.

Then another link with brain damage was discovered. Those who died of Alzheimer's disease, which is also a progressive dementia, were found to have abnormal deposits in their brains called senile plaques, which analysts showed had aluminium silicates in them. For a while aluminium was thought to be the cause of Alzheimer's. but the deposits are now seen as more likely to be a symptom of the disease. If you are still worried about aluminium in your diet then you should read the book by Dr John T. Hughes called *Aluminium and your Health*. Hughes was a neuropathologist to the Oxford Hospitals and the University of Oxford for many years, and at the end of the book he reassures us that: 'Having studied Alzheimer's disease for many years, I am convinced that aluminium plays no part in its causation.'

We cannot avoid aluminium since it is so abundant, and it is rather surprising that it has no metabolic role. The aluminium we ingest normally passes straight through us, and any that gets into our blood is rapidly excreted. However, aluminium can attach itself to a molecule in the blood called transferrin, which is used to carry essential metals about the body, and this is the way it enters the brain.

The average person takes in about 6 mg of aluminium a day (about 2 g a year). The amount depends on what food we eat, whether we cook it in aluminium pans, whether we suffer from indigestion, and whether we prefer tea to coffee. For example, processed cheese has a lot, almost 700 ppm of aluminium. Cakes and biscuits may be made with a leavening agent sodium aluminium phosphate, and aluminium sodium silicate is put into powdery foods to keep them free flowing.

The aluminium which gets into food from pans is small, even when we cook rhubarb. The oxalic acid in this has a curious spin-off, as we saw in Gallery 1. It leaves the pans almost like new because it dissolves the surface layer of metal oxide, a trick our grandmothers knew. Aluminium has a particular affinity for oxalic acid, and for fruit acids such as citric acid, and this combination can make it easier for aluminium to be absorbed by the body. Yet the amount of aluminium in the stewed rhubarb is still tiny. Cooking in aluminium pans generally adds about 1 ppm to food, not enough to worry about.

Tea can be the main source of aluminium in a tea-drinker's diet. Tea plants absorb aluminium from the soil. Indeed, aluminium potassium sulfate is used as a fertilizer for tea plantations. The average cup of tea has about 4 ppm of aluminium, which is 20 times the recommended level, and those who drink strong tea can be imbibing a solution of 10 ppm aluminium.

Although for many years there was a good deal of concern about aluminium in the diet, it now appears that the metal is relatively harmless. We can consume a year's supply of aluminium in one day if we take six aluminium hydroxide indigestion tablets. A pint glass of the Camelford water would have given the drinker as much aluminium as half an indigestion tablet.

There may be other more worrying metals lurking in a glass of water, which is why public supplies are regularly analysed to see that they do not exceed the maximum legal levels, which are generally those set by the World Health Organization—see the Table overleaf.

Drinking water standards

World Health Organization (WHO) 1993 Guideline of maximum acceptable levels of inorganic substances dissolved in drinking water. These level are given in mg per litre (ppm.)

Antimony	0.005	Lead	0.01
Arsenic	0.01	Manganese	0.5
Barium	0.7	Mercury	0.001
Beryllium	n.a.*	Molybdenum	0.07
Boron	0.3	Nickel	0.02
Cadmium	0.003	Nitrate	50
Chromium	0.05	Nitrite	3
Copper	2	Selenium	0.01
Cyanide	0.07	Thallium	n.a.**
Fluoride	1.5	Uranium	n.a.***

* According to the WHO there are no adequate data on which to base a value, but the US Environmental Protection Agency (EPA) sets a value of 0.001.
** not listed by the WHO, but the US EPA sets a value of 0.002.
*** no adequate data on which to base a value, and neither Europe nor the US agencies give a value.

GALLERY 7

WE'RE ON THE ROAD TO NOWHERE

An exhibition of molecules to transport us

■ FUEL FOSSILS ■ GROW YOUR OWN PETROL
■ TURNING COAL INTO PETROL ■ FIELDS OF GOLD ■ CLEAN AND COLD
■ UNDER PRESSURE ■ MAKING THE STREETS SAFE
■ A BREATH OF RED MAGIC ■ SPARE THE TREES
■ BANG! YOU'RE NOT DEAD!

THE RAYS of the Sun, and the motions of the Moon and Earth, provide energy in abundance. Light from the Sun is absorbed by plants on land and algae in the sea and is used to convert carbon dioxide into high energy carbohydrates, which in turn become oils. Together these provide most of the food energy for animals like ourselves. We can also harvest plants and trees and burn them to release this energy as heat. The sunlight which falls on barren terrain, or on the roofs of buildings, we can also gather by using solar panels to heat water or to make electricity. The sunlight which falls on the oceans leads to evaporation of water which is precipitated on land, and this too we can use to generate hydroelectricity.

The Earth itself is a vast reservoir of heat below the crust, but this is not so easily tapped—although in parts of the world, such as New Zealand, hydrothermal heat is an important source of power. We can extract energy from the effects of the Earth's daily rotation, partly through the weather systems this produces, by using windmills, and possibly through the rise and fall of sea levels, by using tidal barriers and wave power. These sources of clean energy should be able to provide all the fuel and electricity for a sustainable human population of several billion, provided we did most of our travelling on foot or by bicycle.

How much these natural renewable sources could really provide is debatable, but we have the means to utilize them so they could supply enough food and energy for a world population of two or three billion, and at a level which allows for most of the high-tech living that we now take for granted. It might even be possible for most families to run a car, provided they were content to travel only a couple of thousand miles a year in it. The trouble is that there are already six billion of us, and forecasts are that this will reach ten billion by the middle of the next century. Most of these people will no doubt aspire to owning a car.

Portrait 1

Fuel fossils—carbon

Until the human population decreases we have no option but to tap the vast reserves of nuclear and fossil fuels with which this planet is endowed. Neither source is inexhaustible, of course, but they will last a long time. Fossil fuels were accumulated over hundreds of millions of years, but they appear likely to be in short supply one day, despite the fact that each year known reserves get larger. They are vast but limited, and perhaps in a way we do not appreciate. Reserves are calculated as those that are economically accessible, in other words cheap to extract. The less economically accessible reserves are also exploitable, but they require more sophisticated techniques for getting at them.

Since we are so dependent on fossil fuels, and are likely to remain so for a long time yet, we need to know how much carbon there is locked away in the Earth's crust, and how much of this we can use. This sounds an almost impossible thing to know, but a little chemistry will give us the answer—or at least *an* answer.

We can calculate roughly how much fossil fuel there is if we look at the amount of oxygen in the atmosphere. Every oxygen molecule (O_2) has been produced from a molecule of carbon dioxide (CO_2) and the carbon it has left behind must be somewhere—generally in coal, oil or gas. There are also tar sands whose extent is vast. Carbon in its reduced states takes the form of natural gas (which is CH_4), oil (which approximates to CH_2), or coal (which has the least hydrogen of all and is roughly CH). When reduced carbon is oxidized by burning, large amounts of energy are released, and the carbon returns as its oxidized form, carbon dioxide (CO_2), back to the atmosphere.

Carbon can exist in a variety of 'states', depending on how many chemical bonds it forms to oxygen or hydrogen. The more oxygen it has, the more oxidized it is; the more hydrogen, the more reduced it is. And the more reduced the carbon is, the more energy is released when it burns and is converted to its oxidized form, CO_2.

Split 100 tons of carbon dioxide into its elements and you will get

73 tons of oxygen gas and 27 tons of carbon. You can calculate this knowing that a molecule of CO_2 has a total weight of 44 units, made up of 12 parts carbon and 32 parts oxygen. There are 1000 trillion tons of oxygen in the Earth's atmosphere, and all this originated from carbon dioxide by the process of photosynthesis (from the action of sunlight and plants). We can calculate that this amount of oxygen must have come from 1,375 trillion tons of carbon dioxide, and that the difference of 375 trillion tons must be carbon somewhere in the Earth's crust.

How much of this do we use each year? The answer is a surprisingly small fraction of only 0.007 trillion tons. This is the same as saying 7 billion tons, and it is calculated each year by adding together the amount of carbon in the natural gas, oil and coal that is used. You can see that at this rate it will take more than 50 000 years to use it all up. This of course we can never do because by then the oxygen in the atmosphere would all be back to CO_2 and all animals long extinct. Humans would have to give up burning fossil fuel long before this because we would find life unpleasant if the amount of oxygen in the air dropped much below its current level of 21%. If it fell to 17% we would be gasping for breath, and below this we would die, as we saw in Gallery 6. Even so, a reckless millennium of fossil fuel burning could lower the oxygen content of the atmosphere—assuming that we could use up the oxygen faster than plants could replenish it, which they might easily do using the bounty of carbon dioxide we would be providing for their increased growth.

Given that we wish to avoid long-term damage to our planet, we should plan to use energy as efficiently as possible. This is where chemists come into the picture. They cannot solve the problem of too many people—that is a religious, cultural and ethical issue—although they can help those who want to enjoy sex but don't want to produce a baby. What chemists can do is improve energy efficiency by designing better materials for lighter vehicles, and better insulation for buildings. They can help the shift to renewable resources by finding better materials for solar panels, and better ways of farming high energy crops. Meanwhile we can try and get the maximum amount of energy from the fossil fuels that we do extract from the Earth. But how much energy do we use? How much energy can we really get from oil, the fossil fuel we currently use the most to transport us around?

A barrel of crude oil ends up as petrol for cars, kerosene for aircraft, diesel fuel for heavy vehicles, lubricants for engines, and bitumen for roads and roofs. About 10% of the oil is also turned into petrochemicals, and these go to producing a myriad of products, some of which we saw in Gallery 5. To bring all this about requires around 5% of the oil in the barrel to be burned to create the energy needed to process the other 95%. How this ends up depends on the source of the oil, the efficiency of the oil company which extracts it, the economic needs of the country in which it is refined, and the demands of customers. This is the complex world of economics. The chemistry is a little easier to understand.

We can classify carbon-based fuels, and what they are used for, in terms of how many carbon atoms a molecule contains. The simplest hydrocarbon is methane (natural gas) with one carbon atom (C_1), followed by ethane with two (C_2), then propane (C_3), butane (C_4), pentane (C_5), hexane (C_6) and so on. These molecules, and those with more carbons in this series, are collectively known as alkanes. The hydrocarbons in gasoline (petrol) are C_5–C_6, while kerosene (paraffin) is C_{10}–C_{14}, diesel is C_{14}–C_{19}, white spirit (the same as the turpentine produced in tree oils) is roughly C_{10}; liquid paraffin is C_{20-25}, and lubricating oils are C_{30-45}. As the alkanes get bigger they become less volatile, and so less dangerous.

When oil is heated, the most volatile components—those with the smallest number of carbon atoms—come off first. These are mainly the light hydrocarbons propane and butane. These gases are liquefied for ease of transportation, and liquid petroleum gas (referred to as LPG) is shipped and used all over the world.

Raise the temperature of the oil, and liquid hydrocarbons distil off next: first gasoline (petrol), which may account for up to 40% of the barrel, then kerosene (paraffin) which is used mainly as aviation fuel; and finally diesel fuel for trucks and, increasingly, for family cars.

To extract other products, the oil has to be heated under vacuum. The products which distil over at this stage can be further processed through a catalytic cracker to convert them to more of the above, the higher value products. Heating oil, lubrication oils and bitumen are then left at the bottom of the barrel.

The more we convert a barrel of oil into light fractions, the better. Yet even the fraction of the oil left behind, the bitumen, is needed, though this may account for only 2%. We normally think of bitumen as just a sticky black tar, but this overlooks a long history of development. Most bitumen is used to make asphalt for roads, binding together the ballast of aggregate, sand and chippings. Although bitumen accounts for only 6% of the asphalt, it is the key to how well it performs. By careful blending, and the use of additives, it is possible to produce bitumen emulsions that can be applied cold to the road surface, or polymer-modified bitumens which are extremely tough and durable, or even clear bitumens that can be coloured to provide permanent road markings or attractive paths and pavements. Modern roads are quieter and throw up less spray during wet weather when so-called 'whispering asphalt' or 'popcorn asphalt' is laid. This has a more open texture leading to less tyre noise and better water drainage, and in some countries this is the only type of asphalt that is now permitted.

Not only have roads to meet environmental criteria, such as less noise pollution and the use of industrial waste for aggregates, but they are also being constructed in regions with extremes of climate. It is possible to modify bitumen by the addition of polystyrene-type polymers, thereby enhancing its adhesive and self-healing features. This improved elasticity, flexibility and crack-resistance makes it ideal for roofing as well.

Fossil fuels may be at the industry end of the energy chain, but what about the other end where consumers use it? How much energy does a typical person consume? The *BP Statistical Review of World Energy* gives a total world consumption of all types of energy as 95 000 billion kilowatt hours (kWh) per year, which divided by the world's population of 5.5 billion gives a total of 17 000 kWh per person per year. Obviously there are many in the world who see only a tiny fraction of this, and many who consume double or triple this amount. This average figure is equivalent to two kilowatts per person per hour, day and night throughout the year.

Another way of looking at the use of energy is to consider a typical household. Assume a family lives in a temperate climate, in a house with three living rooms and three bedrooms, and also runs a family car. This might seem atypical of the world in general, but it approximates to what most families aspire to, and what many have already achieved. We all live in homes of varying efficiency; we don't all have cars, although this is changing rapidly. In North America there are 50 cars for every 100 people, in Australia the figure is 45, in Europe it is around 40, and in Japan 30. The world average is 10, with countries like India and China having fewer than 1. We might expect there to be one car per six people in a world where the population was stable, and the average family consisted of two adults, two children and two aged parents. This ratio of 17 cars per 100 people would still mean a thousand million cars for the current world population.

Whatever people aspire to, we can nevertheless calculate the energy our typical family might use. The figures given in the table on the next page are measured in kilowatt hours (kWh) over a whole year, and are based on information published by the oil company Shell, and by the UK's Department of the Environment and Central Office of Information.

Energy used by a typical family in a developed economy.

Energy used	kWh	Energy %	Cost %
Heating (gas)	13 000	45	18.5
Hot water (gas)	4 500	16	6.5
Car (petrol)	8 500	30	56
Cooking (electricity)	1 000	3	7
Fridge/freezer (electricity)	600	2	4
Dishwasher (electricity)	500	2	3.5
Lighting (electricity)	250	1	2
Washer/drier (electricity)	200	0.5	1.5
TV, etc. (electricity)	150	0.5	1
Totals	28 700	100%	100%

The table is concerned only with the energy that we as individuals have control over, in our homes and cars (tax on petrol explains the high percentage cost of running a car in most countries). It does not include the vast amounts of energy used in work places, public buildings, public transport, air travel and trucks. Warmth and warm water we can hardly do without, but our use of the family car is something we could well consider more critically.

Perhaps our love affair with the motor car will never end. They are easy to fall in love with, offering some very seductive pleasures. They give us mobility for ourselves and our possessions, in comfortable surroundings that protect us from the elements; and they can even entertain us with music or radio. Some day this love affair will have to end unless we can develop sustainable sources of energy to replace those derived from fossil fuels, which currently drive 95% of the world's vehicles. We can already see what these alternative fuels for transport are likely to be, and how they might be derived from renewable resources.

There are more than 600 million cars in the world, and these need a liquid fuel to power their engines. Most run on petrol, and a growing number are being produced that run on diesel. Could the fuel to run them come from renewable resources, and might it be possible to find one that, unlike current fuels, does not pollute the atmosphere?

Liquid hydrocarbons make excellent liquid fuels and are used for cars, lorries and aircraft. They release a lot of energy when they are burned—about 33 000 kJ per litre (It takes about 400 kJ to boil a litre, or a quart, of water). Whatever we replace them with must also pack a lot of energy. The main contenders which come from renewable sources, the biofuels, are ethanol, methanol and rape methyl ester.

'Biofuel' is the general term given to fuels that are regarded as sustainable because they can be harvested. Some crops give particularly high yields of oils, such as the gopher plant (*Euphorbia lathyris*) which oozes a milky-white latex sap. In theory this could be used as feedstock for the oil industry. The Brazilian tree *Cobaifera langsdorfii* has a sap which can be put directly into the fuel tank of a diesel engine. As yet neither plant has been developed as a source of fuel, but it might one day be possible through genetic engineering to increase the production of oil by trees, and even make them withstand temperate climates so that each kitchen garden could have a tree slowly dripping biofuel for the family's diesel-powered car.

Much more likely will be the industrial production of fuels we already know. So how much energy can we get from typical fuels? The table below lists some that are already used, and others that are likely contenders for the future. The information comes from *Macmillan's Chemical and Physical Data* (1992). The figures in the table refer to kJ of energy per kilogram of fuel, rather than kJ per litre, because both gaseous and liquid fuels are included. What is important about a fuel is the energy it releases when we burn it. The higher this is the better, because a car will have to carry less weight as fuel which means it will be able to travel further on a tankful. Some of the fuels listed in the table are gases, but the technology for using these is already in place. If we are to look at oxygen-containing chemicals like ethanol and methanol, then clearly, because they are already partly oxidized, they will release less energy per litre.

The energy of fuels.

	kJ/kg
Hydrogen gas	143 000
Methane gas	56 000
Petrol, kerosine, diesel	48 000
Rape methyl ester	*c.* 45 000
Ethanol	30 000
Methanol	23 000

There is another factor to be taken into account when considering the change from one fuel to another: how safe is it to use? We are familiar with petrol, and there are surprisingly few accidents when people fill up their cars. Yet it is highly combustible and therefore potentially dangerous, especially if the tank is ruptured in an accident. If an alternative fuel poses no additional hazards, then we can have no objections to it.

Methanol is safer than petrol in terms of flammability, and while it is safer in a crash it is harsh on engines. In addition it is a toxic material, and so poses a health hazard. Ethanol appears to offer the best combination of convenience and safety of any liquid fuel. Biodiesel presents no special difficulty. Liquefied gases, on the other hand, do bring new types of hazards especially during refuelling, which has to be done more often than with petrol.

Portrait 2

Grow your own petrol—ethanol

The biofuel which has been most successful is ethanol, popularly known as ethyl alcohol or just alcohol. Ethanol is good as an alternative fuel for cars, despite its slight disadvantage of providing less energy than petrol (gasoline). A litre of ethanol releases 24 000 kJ of energy when it burns, compared to the 33 000 kJ which can be released by burning a pure hydrocarbon like petrol.

In Brazil, Zimbabwe and the USA a large amount of biofuel ethanol is produced from sugarcane and surplus grain. The Brazilians saw ethanol as their way of overcoming the growing burden of costly imported oil in the 1970s and 80s, and although they now have their own indigenous sources of fossil fuel oil, they still produce about 12 billion litres of ethanol a year. Brazil still has millions of cars running on ethanol, while millions of others use a mixture of 20% ethanol and 80% petrol. Ethanol accounts for more than 20% of Brazil's transportation fuel, although this has declined from a peak of 28% in 1989. The cost of ethanol is no longer competitive, despite the efforts of farmers who have increased sugarcane yields by 20% to 77 tons per hectare, and despite manufacturers boosting alcohol production with improved methods of fermentation and distillation. Ethanol is still popular because it has other advantages as a so-called oxygenated fuel, and this means it produces less pollution when it burns. Ethanol enabled Brazil to be one of the first countries to phase out leaded petrol, to cut the levels of carbon monoxide in the air of their cities, and to banish photochemical smog caused by unburnt hydrocarbons. This can be seen in São Paulo, with its 15 million inhabitants, where the air is relatively clear.

The USA produces over 3.5 billion litres of ethanol a year by fermenting the starch of surplus corn and grain, and plans to double this by the year 2000. They too will use it in reformulated blends of gasoline for cities where pollution is a problem. Countries which cannot grow sugarcane can produce sugarbeet which is just as high-yielding, so in theory ethanol could be produced as a renewable fuel in most countries of the world.

Could a Western country produce enough ethanol to support, say, one car per family? The answer would appear to be yes. The average family car travels about 13 km per litre, consuming each year about 1250 litres of petrol, which could be replaced by 1730 litres of ethanol. The requirement for a city with, say, a million cars would be 1730 million litres. The city region would need to plant around 45 000 hectares of sugarcane or sugar beet to meet its transport fuel needs, and this would cover an area approximately half as big again as the city itself. If the city had a radius of 10 kilometres then it would need a strip of land about six kilometres deep surrounding it to grow the ethanol it needed for its cars. Clearly, it can be done.

Portrait 3

Turning coal into petrol—methanol

Methanol used to be produced simply by heating wood chippings in a retort, and for this reason it was called wood alcohol. Its other name is methyl alcohol. Today it is made from synthesis gas (syngas), which is the gas that forms when steam reacts with carbon, such as coke or coal, or hydrocarbons, such as oil or natural gas. Syngas can be turned into methanol using a zinc oxide/chromium oxide catalyst. Methanol in its turn can be converted into petrol by passing its vapour through catalysts made of zeolites, which are aluminium silicates with large open cavities. The methanol diffuses into these cavities and emerges as petrol (gasoline).

Syngas is seen as opening up a newer and cleaner age for coal and oil residues. Turning these into syngas means they can be burned to generate electrical energy without the emission of sulfur dioxide and other polluting gases. At Buggemun in The Netherlands is the world's largest coal gasification plant, which turns 2000 tons a day of coal into the syngas which generates 250 megawatts of power. The flue gas is filtered through ceramic traps to remove any particulates, while the molten slag is turned into ballast for road construction.

Converting coal to oil began with the Fischer-Tropsch process in Germany in the 1920s. Syngas from coal was passed over supports which held iron or cobalt catalysts with promoters and was thereby converted to more useful liquid hydrocarbon. The technology enabled poor quality coal to be turned into petrol, diesel and aviation fuel and provided the Third Reich with 600,000 tons of liquid fuel per year in World War II. In 1944 the producing plants were heavily bombed in a concerted attack on German fuel supplies. Production virtually ceased and this denied the Luftwaffe its aviation fuel, thereby hastening the end of the war in Europe.

Fischer-Tropsch technology also supported the South African economy in the 1970s and 1980s when its fuel supplies were cut off in the world-wide embargo against apartheid.

Currently Fischer-Tropsch technology is in operation at the Bintulu, Sarawak, plant in Malaysia where it is producing top-grade hydrocarbons which are free of the sulfur and nitrogen compounds that contribute to air pollution when oil-based fuels are burnt.

Syngas can also be turned into methanol, which is a liquid that boils at 65 Celsius, and can be used by itself as a fuel in car engines. Indeed, it is the fuel of choice for drivers in the famous Indianapolis 500 races because it is clean-burning and, unlike petrol, it does not produce a fireball when a ruptured fuel tank catches fire in a crash.

World demand for methanol now exceeds 20 million tons per year, most of which is added to unleaded petrol. This has around 5% of methanol in its formulation. Alternatively, methanol is used to make methyl tertiary butyl ether (MTBE), which is increasingly being added to petrol to reduce pollution by making the fuel burn more cleanly. Such a fuel with added MTBE is said to be oxygenated.

When petrol burns in a car engine it does not all end up as carbon dioxide and water. Incomplete combustion produces pollutants such as hydrocarbons and carbon monoxide. The unburnt fuel then is acted upon by sunlight to form more unpleasant pollutants. To combat this, the US Government passed the 1990 Clean Air Act Amendment, which decreed that oxygenated fuels must be used at certain times of year. Each winter 29 US cities convert to petrol containing 15% of MTBE. More than six billion litres of MTBE are produced each year for this purpose, and production continues to rise.

MTBE has not been without its opponents. Some say it increases the levels of other pollutants, such as formaldehyde, in the exhaust gases, while others accuse MTBE itself of being a health risk, causing headaches, dizziness, sore eyes and nausea. This was not borne out in comparative studies involving garage and forecourt workers in northern New Jersey, where MBTE is used, and those in southern New Jersey, where MBTE is not needed. A 1996 report by the US National Research Council also showed that concerns about the safety of petrol additives was unfounded. There were no differences in symptoms or complaints between the two groups. The unpleasant smell of MTBE is thought to be the cause of the complaints. Extensive tests on rodents showed that this chemical can be tolerated at high levels, and any that is absorbed into the body is rapidly converted to tertiary-butyl alcohol and eliminated in the urine.

Methanol looks set to play an increasing role as a fuel additive, chiefly as MBTE. In theory, it would be possible to generate methanol from renewable resources such as wood charcoal, which could be the carbon source for making the syngas. Methanol may also be used more cleanly in fuel cells to generate electricity for an electric-powered car, but that technology is still a long way off.

Portrait 4

Fields of gold—rape methyl ester, RME

The patchwork of bright yellow fields in many countries proclaims the popularity of rapeseed, a crop which has one of the highest yields of vegetable oil and which can be grown almost anywhere. In Europe rape will grow as far south as the Mediterranean and as far north as Scotland. Rape makes an ideal 'break' crop that farmers can plant between runs of cereal crops to interrupt the pattern of cereal diseases.

Rape is a member of the mustard family. Its scientific name is *Brassica napus*, but its common name comes from *rapa*, the Latin for turnip—the variety grown for its enlarged edible root. In the northern hemisphere rape is either sown as winter rape in September, or as spring rape in early March, and it is harvested in early and late June respectively. Winter rape yields about four tons of seed per hectare; spring rape yields about three. A ton (1000 kg) of rapeseed produces about 320 kg of oil when crushed, but the bulk of the crop is pressed cake. This is high in protein and is used as animal feed, but it can cause thyroid disorders because of high levels of compounds called glucosinolates.

Oilseed rape had been economically very important in the 400 years since it was developed as a hybrid between cabbage and turnip. In 1572 the English Parliament passed a bill to encourage the development of a local oil-yielding crop 'equal to Spanish and foreign oils.' By the end of the next century there was a thriving rapeseed oil industry producing oil for lamps and meal for cattle feed. This industry continued for over 200 years until the introduction of, first, whale oil and then fossil fuel oil in the late nineteenth century, whereafter it went into rapid decline.

Demand for its products, which were used in certain industries, were then met from China and India where it was still widely grown. The World Wars of the twentieth century again increased demand for rapeseed oil, particularly as a marine lubricant. Today it is flourishing again—as a new variety.

The original rape contains a fatty acid known as erucic acid. This has 22 carbon atoms in its chain, and it became a major health concern in the 1970s when research in The Netherlands and Canada showed that a diet rich in erucic acid caused excess fat deposits to form in the heart muscle of young mammals. Consequently plant breeders selected seeds which had low levels of erucic acid and produced a variety of rape, known as canola, which is not only low in erucic acid but low in saturated fatty acids as well. In China and India high-erucic rapeseed oil is still widely available and used.

Erucic acid is the oil which was featured in the harrowing film *Lorenzo's Oil*, about a woman's struggle to save the life of her son using a treatment which involved feeding him high doses of erucic acid. This had to be obtained from a British company, called Croda, that processed rapeseed oil and was able to extract and purify the acid for Lorenzo's treatment.

Rape is a valuable source of edible oil, but it could also be a renewable resource of hydrocarbon oils for the chemicals industry. In the old days rape was used to make components for plastics, synthetic rubber, soap and lubricants, or it was burnt as heating oil.

Vegetable oils (and animal fats) are molecules called triacylglycerides. They consist of three long-chain fatty acids connected to the small molecule glycerine. Rapeseed oil can be burnt in diesel engines without further modification, but the engines tend to become clogged with glycerine compounds after a few days. If the oil is broken down into its fatty acids, by heating with alkali, and these are treated with methanol at 50 Celsius, the glycerine settles out and the biodiesel can be separated off. Biodiesel is more correctly called RME, which is short for rape methyl ester.

In Europe RME is made by the Italian firm Novamont, which supplies it for public transport in 17 Italian cities, for taxis in Berlin and Bologna, and for the Lake Como ferry in Northern Italy. In Austria RME is sold at 100 filling stations, and several thousand hectares of rape are grown there to supply them. RME plants are planned or are being built in several European countries, and by the year 2000 there could be over 500 000 tons a year of RME being manufactured. Untaxed RME costs about the same as diesel which is taxed. Even if it were taxed, RME might still be preferred for buses and taxis in city traffic because, unlike diesel, it emits no sulphur dioxide, and it produces fewer sooty particulates.

Most rapeseed oil will probably continue to be used for human consumption. Currently 90% of the oil ends up in foodstuffs, particularly margarine, followed by cooking fats, biscuits, crisps, soups, ice cream and confectionery. The industrial users, who take the remaining 10%, are pharmaceutical companies, which culture bacteria on it, and the chemicals industry, which uses it for plastic bags and cling film.

Rapeseed oil grown for food purposes has fatty acids which are 6% saturated, 64% monounsaturated and 30% polyunsaturated. A chain is said to be saturated when each of its linking carbon atoms has two hydrogens attached. If hydrogen atoms on adjacent carbons are missing the chain is said to have an unsaturated bond. If the chain has one such bond it is called monounsaturated, and if it has two or more such bonds it is called polyunsaturated. Some food advisors claim that monounsaturated fats are the healthiest, and so rape scores highly—almost as highly as olive oil, which is 77% monounsaturated. Rapeseed oil may contain small amounts of erucic acid, which is also monounsaturated but gives the oil an unpalatable taste if there is too much of it. The older varieties of *Brassica napus* contained up to 50% of this acid but selective breeding led to the varieties we now grow, and these have less than 1% of erucic acid. There are moves to reintroduce the older varieties because erucic acid has a market with the chemicals industry, and crops of this type of rape could be grown on farm land no longer needed to grow food crops. It too could be treated with methanol to produce RME for diesel engines.

In theory any vegetable or animal fats could also be turned into biodiesel, and a few years ago this was proposed as a way of using up unwanted tallow from sheep in New Zealand. The buses which serve the Logan airport in Boston, Mass., have been burning a fuel that contains 20% waste animal and vegetable fat that has been collected in the city.

Algae can also make oil and, given the right conditions, they can double their numbers five times a day. Not only that, but some algae will convert over half their mass to oil, and they grow well in polluted water—and even in water that is saltier than seawater. In a sunny climate a hectare-sized pool (10 000 square metres) could produce 120 tons of algae a year, more than double the biomass yield from a crop such as rapeseed or sugarcane. Paul Roessler of the Solar Energy Research Institute (SERI) at Golden, Colorado, claims that such a pond could yield almost 100 000 litres of fuel per year. Again, the oils would be converted to their methyl esters. SERI are investigating the diatoms *Chaetoceros* and *Navicula* and the green algae *Monoraphidium* as possible candidates for algal fuel.

Meanwhile, studies at the Research Institute for Agrobiology and Soil Fertility in The Netherlands have shown that it should be possible to increase the yield of rapeseed oil from 2.5 tons per hectare to over 5 tons by a combination of early flowering, late ripening and clustered pods. These features are already present in various varieties of rape but now need to be brought together in the same plant.

The trouble with all biofuels is that they tend to have hidden costs; in other words, we may produce a litre of ethanol, methanol or RME and assume it is all clear profit in energy terms. But it isn't. A large amount of other kinds of energy has gone into producing it. Crops need planting, fertilizing, protecting, harvesting and processing before we have a usable fuel, and all these steps use energy. Indeed, it has been calculated that to produce a litre of RME, about a litre of fossil fuel has to be consumed. Clearly the exercise of growing biofuels would be is pointless if this were the case, but there may be ways in which the crop can be produced without the need for fossil fuels. Perhaps it would need only an input of electricity which might be generated at a hydroelectric plant, but to construct this there has to be the production of large amounts of cement and that too requires a great deal of fossil fuel.

Some day our descendants will have to find ways of living without fossil fuels, so these problems will have to be solved. Meanwhile it might be better if we looked more closely at other molecules which could be produced from renewable sources.

Portrait 5

Clean and cold—hydrogen

James Dewar (1842–1923) invented the vacuum flask, which we associate with keeping hot drinks hot. But Dewar used it to keep very cold liquids very cold. It was thanks to his silvered, glass vessels with their vacuum layer that in 1898 he was able to liquefy hydrogen gas for the first time, and, the following year, to produce solid hydrogen. Hydrogen melts at −259 Celsius and boils at −253 Celsius. It is the lightest of all gases, and it was first made in 1766 by Henry Cavendish, by reacting metals with acids. (We met this remarkable man earlier in this Gallery.)

Throughout its history hydrogen has had links with transport—but always in rather impracticable ways. The same is true today. Hydrogen gas was first used by Henri Giffard to raise an airship in Paris in 1852, but it was in this century that so-called 'dirigibles' had a brief but spectacular career. They were used to bomb London and Paris in World War I, and ran passenger services across the Atlantic in the 1920s and 30s, until the spectacular *Hindenberg* disaster in New Jersey in 1937 when this airship exploded while coming in to land, with the loss of many lives.

More recently hydrogen has been advocated as a fuel for cars and Daimler-Benz Aerospace Airbus is developing a hydrogen powered aeroplane. The problem of hydrogen is storing it, because hydrogen takes up a lot of space when it is a gas. One kilogram of gas occupies 11 cubic metres, which is 11 000 litres. Condensed into liquid hydrogen, however, it occupies only 14 litres, and delivers three times as much energy as the same volume of petrol. Liquid hydrogen technology already exists, and is needed to handle the enormous quantities required by the US space programme, which uses road and rail tankers to move 75 000 litres at a time. One storage tank at Cape Canaveral holds over three million litres of liquid hydrogen.

There are plans to supply hydrogen as a fuel to homes once natural gas supplies run out. Because hydrogen is a lighter gas than methane it requires three times the volume of gas to provide the same amount of heat. There are environmental benefits to offset this disadvantage: when it burns, hydrogen produces only water vapour. One day there may be a 'hydrogen economy' in which this gas will be piped into our homes for heating and cooking, and maybe even for refuelling the family car.

Hydrogen cars have been demonstrated in Japan. A team of scientists led by Shoichi Furuhama at the Musashi Institute of Technology has worked on a hydrogen car for over 20 years, and in 1992 their car completed its trials, running 300 kilometres (200 miles) on a 100 litre tankful of liquid hydrogen. The fuel is held in a stainless steel version of a Dewar vacuum flask. The Musashi car is a Nissan Fairlady Z sports car in which the diesel engine has been modified with a spark plug so that it will ignite hydrogen gas under 100 atmospheres pressure. BMW in Bavaria in 1996 had six hydrogen-fuelled cars on the road, and is a world leader in their field. The company expects to start marketing hydrogen-fuelled cars in 2010, and that by 2025 about 2% of cars will run on this fuel. Robot filling stations will be developed to recharge a car with liquid hydrogen, and will be able to do this in three minutes with zero loss of fuel. (Filling a normal car loses around 2% of the fuel.)

Currently hydrogen costs three times as much as petrol, and the car itself will cost about twice as much in any case. Liquid hydrogen is carried in a 120 litre cylindrical fuel tank under a pressure of five times atmospheric and kept cool by 70 thin layers of insulating aluminium foil and glass fibre which will fill a 3 cm gap between the tank and the outer container. A full tank weighs about 60 kg (27 lbs) and is enough to enable a medium-sized passenger car to travel 400 km (250 miles). The hazards are no greater than those in a petrol-powered car, but the distance travelled is only about half as far for the same weight of fuel.

One source of Europe's hydrogen could be Canada. There the Euro-Quebec Hydro-Hydrogen Pilot Project has shown that cheap hydropower from the St Lawrence River could produce the gas on a large scale, and this could be shipped across the Atlantic in 200-metre-long tankers each holding 15 000 cubic metres (15 million litres) of liquid hydrogen.

But hydrogen as a fuel need not be carried round as a liquid: it can be absorbed and stored in certain metal alloys. Mazda have built a hydrogen car which they exhibited at the Tokyo Motor Show in 1991, and which stores its fuel this way. Alloys of titanium and iron, or magnesium and nickel, can absorb the equivalent of their own volume of liquid hydrogen and release it as required. Inside the alloy the hydrogen packs into the spaces between the atoms of metal. In the Mazda car hydrogen is not burnt, but is used to generate electricity in a fuel cell where the hydrogen releases its electrons to produce an electric current and then combines with oxygen to form water.

Unfortunately the storage of hydrogen in alloys is beset with difficulties. Pumping hydrogen in and out of the alloy makes the metal brittle and reduces it to dust after a time, and if traces of moisture get into the storage tank its capacity is much reduced. Problems like these, and the high cost of hydrogen, make it impractical as a fuel for cars, and it is unlikely that hydrogen cars will ever be popular.

If there is ever to be a hydrogen economy it will clearly need a lot of gas. Hydrogen is already manufactured on a large scale for the chemicals industry, and is piped over hundreds of kilometres around Europe and the USA. It is used in many ways, but most goes into making ammonia, hydrogen peroxide and margarine. A lot of hydrogen gas is generated as a co-product in the manufacture of sodium hydroxide (caustic soda). The gas is either burnt to generate electricity, or piped to other companies, for example to make hydrogen peroxide, or it is sold in smaller lots in high pressure cylinders. A truck stacked with cylinders demonstrates the economic difficulties of hydrogen: a fully loaded forty-ton lorry is actually transporting less than half a ton of the gas and a mere fraction of the energy of a loaded petrol tanker.

World production of hydrogen is around 350 billion cubic metres per year, which is about 30 million tons. There are two natural sources of hydrogen: water (H_2O), and hydrocarbons like methane (CH_4). There is also an estimated 130 million tons of hydrogen in the Earth's atmosphere, but this is too dilute to be reclaimed, and it is being slowly lost to space.

Hydrogen is released from water by passing an electric current through it, but this is not an economical process despite several improvements in efficiency, such as the electrolysis of steam inside porous electrodes of zirconium oxide. Using the surplus electricity of hydroelectric power stations at night to generate hydrogen would be one way to make the gas cheaply.

Another way of generating renewable hydrogen is to make syngas from renewable charcoal, but this produces a mixture of hydrogen and carbon monoxide gas which is better converted into the alternative fuel methanol, as we have seen. A third method, which is still only a research novelty, is to use sunlight to split water into its component gases, oxygen and hydrogen. Powdered titanium dioxide doped with platinum metal was discovered to do this twenty years ago, but the amount of hydrogen given off was tiny. Kazuhiro Sayama and Hironori Arakawa of the National Chemical Laboratory for Industry at Iaraki, Japan, showed that the yield could be greatly increased by adding washing soda (sodium carbonate) to the water, but even so we are still a long way from a commercially viable way of generating hydrogen from sunlight.

Finally, there are some heat-loving bacteria that are able to produce hydrogen gas themselves. In 1996 a group of scientists at the Oak Ridge National Laboratory in Tennessee showed that glucose dehydrogenase from *Thermoplasma acidophilum*, which was discovered in smouldering coal slag heaps, and hydrogenase from *Pyrococcus furiosus*, which comes from volcanic vents deep in the Pacific Ocean, were together able to generate hydrogen gas from glucose molecules, converting it into gluconic acid. Both enzymes are resistant to heat and so can work at high temperatures, making the process quicker. Given the large amounts of cellulose produced as biomass each year, and that this is a polymer of glucose, it may one day be possible to make hydrogen from wood and waste paper by using such enzymes.

Portrait 6

Under pressure—methane

Ethanol, methanol, RME and hydrogen offer possible sustainable energy sources with which we could power our cars. Were these fuels not enough to supply all that we need, there are other sources we could tap, such as waste. A lot of energy can be reclaimed by burning municipal waste as fuel in incinerators.

Mehdi Taghiei of the University of Kentucky estimates that the USA could generate 80 million barrels of oil from its plastic waste each year, and has shown that mixing this with tetralin, a high-boiling hydrocarbon solvent, and heating under a pressure of hydrogen gas at 450 Celsius for an hour, will convert 90% of the plastic to a light oil, which, unlike ordinary crude oil, is free of sulphur and therefore easier to refine. Other chemical companies have found better ways of doing this by using a hot, fluidized sand bed, and this will turn all types of plastic, including PVC, into useful hydrocarbons for recycling.

Heat and pressure will convert the cellulose from sawdust and newspapers into oil, in effect subjecting it to the process that produced fossil fuels from ancient vegetation deep within the Earth's crust. When sewage is digested in the absence of air, the product is methane gas which is used in Europe as a fuel to generate electricity. However, this is only commercially viable when supported by subsidies. The methane could also be turned into synthesis gas and thence into methanol. Sewage itself can also be turned directly into oil, a process that was shown to be feasible in 1987 by the Battelle Pacific Northwest Laboratories in the USA. In this process sewage sludge is made alkaline and then heated under pressure. This converts the organic material to crude oil, water and carbon dioxide.

Rather than try to convert wastes to liquid fuels, it might be better to allow them to be worked on by methane-producing microbes, as happens in old land-fill sites. The methane given off from tanks of rotting waste and sewage could be collected and even used to drive cars. Currently such cars run on the methane of natural gas. This can be stored and used as CNG (compressed natural gas) and LNG (liquid natural gas). These are already used as a fuel for transport, and supplied in cylinders. Around the world, half a million vehicles run on CNG, mainly in Italy, Canada and New Zealand. Converting a car to use CNG is easy because, like petrol, it burns well in spark ignition engines, but the fuel tank is heavy and bulky, the consequences of which are reduced luggage space and shorter distances travelled between refuelling stops.

Methane is also a renewable resource and is a major by-product of sewage, and indeed of any organic matter that decomposes under the influence of anaerobic bacteria. Methane can be extracted from the decaying matter in rubbish tips. Both sources could be used to produce CNG after treatment, which would be needed because the quality of the gas recovered is poor since it is mixed with carbon dioxide and nitrogen which have no energy value. While the methane could be separated and sold as CNG, it is much easier just to burn effluent gas to produce heat or run turbines to generate electricity.

Before we go over to using CNG and LNG on a large scale we must be sure that we can handle these fuels safely. The technology for handling them is highly sophisticated and reliable, but we must never forget that they can be dangerous, as can all volatile hydrocarbons. History has shown this to be the case on several occasions. A particularly spectacular disaster happened at Guadalajara, Mexico's second city, on 22 April 1992, when a series of 20 explosions in the sewers demolished scores of buildings and killed 194 people. Residents had complained of the smell of petrol seeping from the drains the day before the accident, but no action was taken. Officials of the state-owned oil company, Pemex, blamed a leak of the alkane, hexane, from a local cooking-oil factory, while others claimed that tons of petrol must have been involved.

In fact it needs only 65 milligrams of an alkane vapour in a litre of air to cause an explosion. The contents of a single 60 gallon drum would have been enough to blow up a kilometre of sewer pipe two metres in diameter. The repeated explosions are explained by the behaviour of an air–vapour mix. This would burn along the sewer pipe if the amount of hydrocarbon was low, but would explode every time the flame front came to a richer blend or to a bend in the pipe.

In 1974 a chemical plant at Flixborough, England, exploded when a large tank of hexane caught fire. In 1987 a holiday camp at San Carlos, Spain, became a fireball when a tanker of liquefied propane gas (LPG) crashed. In 1989 a ruptured methane pipeline running alongside the track of the trans-Siberian railway burst into flames and engulfed two passing trains. And in 1984 the worst accident of all happened in Mexico City, when an LPG storage depot exploded killing 542 people and leaving more than 4000 others with serious burns.

Whether such incidents are a thing of the past remains to be seen, but there are now good reasons to believe that we can use methane safely. The gas can be condensed into a liquid which boils at −162 Celsius, and as such it occupies only 0.2% of its original volume. In this manner over 75 billion cubic metres of gas are shipped around the world each year, in special tankers that carry 25 million litres of LNG at a time.

LNG needs to be stored and handled at −160 Celsius, which in hot countries such as the Gulf states can mean 240 degrees below the outside surface temperature. In Oman there are tanks holding 120 *million* litres of LNG sitting in the searing heat. They are a tribute to the skills of engineers, and even if the refrigeration units were to fail, these storage tanks would not burst—the natural gas would slowly evaporate over the next 20 days. In Brunei there is an LNG system that was incorporated in 1971 and which has remained cold and performed safely for over 25 years.

Despite this, we are still haunted by the nightmare of liquefied natural gas explosions, but they are much less likely to happen thanks to research into the way that methane gas behaves. This gas will burn fiercely but it does not explode unless some rather special conditions come together. This can happen if an escape of gas starts to burn, then more fuel is added, the volume expands and it becomes turbulent. It may then burn with explosive violence.

Research into exploding alkane vapour has used lasers, microsecond cine film, and computer image analysis of the colour and shape of the expanding flame front. This has revealed that the disaster hinges on turbulence, which may promote the thorough mixing of air and vapour. If this can be prevented, so can an explosion. Older designs of oil storage installations, oil rigs and chemical plants had a network of pipes, stairways, gangways and supports that actually encouraged turbulence, and this is what happened in some of the major disasters, such as Flixborough. Today designers use better layouts to prevent turbulence.

Test rigs have been built to research the behaviour of massive alkane leaks. Computers can now predict what will happen in a given situation. The rate at which an alkane vaporizes, the wind speed, the profile of the terrain and the proximity of other storage tanks can all be taken into account and the best course of action suggested. Computer programmes derived from this research are now available to plant designers and safety experts worldwide.

But alkanes threaten us in another less dramatic, but potentially more dangerous way. All are very good greenhouse gases, better even than carbon dioxide. Methane is vented to the atmosphere by humans, cows and paddy fields. Leaks of natural gas from wells, pipelines and storage tanks also add a daily quota. Half of Russia's natural gas production is believed to be lost this way. Whether methane can be used on a large scale without adding to this planetary burden remains to be seen.

Portrait 7

Making the streets safe—benzene

Whatever fuel we put into the tank of the family car, it will need other things adding to it if we are to burn it efficiently and protect the engine.

One such additive, introduced in the 1920s, was tetraethyl lead, which was good for engines but harmful for humans. When leaded petrol is phased out, something else has to be phased in to improve the burning efficiency of the petrol needed by older cars that were built to run on leaded petrol. A special lead-free petrol was designed for these cars, and this had benzene added to boost the efficiency. But benzene is also a pollutant. Motorists with older cars could chose either to pollute with lead or with benzene—the choice was theirs.

Other drivers using lead-free petrol need not feel too smug, because most petrol contains benzene: leaded petrol has about 2%, super-unleaded has 5%, and even unleaded contains a little. The extra benzene in super-unleaded boosts the performance to that of leaded petrol, which makes it the green alternative for cars which were built to run on leaded petrol. But few motorists chose super-unleaded, which never accounted for more than a few per cent of sales.

The fuel against which all others are measured is the liquid hydrocarbon iso-octane (a C_8 hydrocarbon), which gives its name to the so-called octane rating. Iso-octane rates 100. When oil is refined into petrol we get a mixture of hydrocarbons with an average octane number less than this. Unleaded petrol is octane 95, and while this is perfectly adequate for modern engines it is no good for older ones. They need the level boosting to 98, and this can be done either by adding a few drops of tetramethyl lead (which is preferred to tetraethyl lead) at the rate of 0.15 g per litre, or by refining the oil so that it has more aromatic hydrocarbons, such as benzene. Aromatics also have high octane numbers, and around 5% of benzene in the blend will give the 98 octane super-unleaded.

Benzene was first isolated in 1825 by Michael Faraday at the Royal Institution in London, when he found it in the gas given off when whale oil was heated. Coal tar provided the vast amounts of benzene that were once used in printing inks, quick-drying paints, dry-cleaning fluids, and rubber solutions for waterproofing fabrics. Benzene is still used to make polystyrene, dyes and nylon, but now it comes from oil.

Benzene is hazardous because it is highly flammable, and breathing its vapour in a confined space can kill. Yet it was once a common industrial and household solvent. Benzene also causes cancer in laboratory animals, and groups of workers exposed to high levels have shown higher incidences of leukemia. Even before World War II it was suspected of causing leukemia in a few people heavily exposed to it. Donald Hunter, in his book *The Diseases of Occupations*, says that by 1939 there had been 14 such cases among the tens of thousands of workers who were in daily contact with benzene. (Ironically, earlier in the century doctors even injected benzene into bone as a treatment for leukemia.) It is the link with leukemia which raises doubts about the wisdom of having any benzene in petrol. Safety regulations proposed for the USA could limit benzene to 0.1 parts per million (ppm), much lower than the current UK standards which permit 5 ppm. But government experts already recommend that benzene should not exceed an average of 5 ppb, which is a thousand times less than current permitted levels, and that even this should eventually be reduced to 1 ppb.

In earlier times 10 ppm and even 100 ppm were considered safe. In the open air the levels of benzene are thousands of times less than this, and are measured in parts per billion. On a cold December day in London in 1991 the level of benzene in air reached a high of 13 ppb, but it is generally only about 2 ppb. Some comes from garage forecourts and some from leaking petrol-tank seals, but most escapes because of incomplete combustion in cars not equipped with catalytic convertors. The level should fall because there is less benzene in petrol today due to improved refining techniques. In the USA the maximum permitted level of benzene in petrol is 1%, and Mobil has developed a refinery process that removes all benzene. Moreover, there are ways of boosting octane other than with benzene, such as with oxygenated molecules like alcohols or ethers.

Another way of reducing public exposure to benzene is to redesign petrol pumps. There are three ways this can be done. The first is to put a bellows round the pump nozzle to catch the fumes from the tank which escape as we fill it, and then return them. A second method proved unpopular with motorists when tried in the USA but is still used in Los Angeles, and this involves sucking the vapour out of the tank as it fills and returning it to the garage owner's underground storage tank. A third method is to absorb the vapour in a charcoal-containing canister in the boot of the car, and then to return this to be burnt when the car is running. Charcoal absorbs benzene vapour.

The amount of benzene we take in through our lungs is the single largest input of this chemical into our bodies, exceeding that from our food and drink. Cigarette smoke also contains benzene. The amounts are so small as to pose no extra risk. In 1990 benzene was detected in Perrier water and this led to a health scare. The benzene came from the carbon dioxide gas used to carbonate the water. Even so, the levels were like those of London's air, about 13 ppb. The amount is so tiny that you would have needed to drink a bottle of such water every day for 100 years to take in half a gram of benzene. In any case Perrier recalled all its existing stocks, dealt quickly with the problem, and issued benzene-free water soon after. Today it is as 'pure' as it has always been.

Portrait 8

A breath of red magic—cerium

Cerium, a little-known metal discovered almost 200 years ago, could be the answer to another environmental problem related to transport. Cerium oxide can eliminate 90% of particulate emission from the exhausts of diesel engines.

Some health experts believe that exhaust particulates pose a growing threat to those who live near busy roads, and emissions of this kind of dust are increasing due to the growing number of cars with diesel engines. These now account for over 20% of new car sales in some countries, and, like buses, vans, lorries and taxis, they emit fumes of fine carbon particles less than a thousandth of a millimetre in size. A diesel car releases about 330 mg of particulates for each mile it travels, whereas a petrol-driven motor emits only 25 mg.

Particulates can lodge in the lungs, and are blamed for an upsurge of respiratory diseases such as asthma and bronchitis. The International Agency for Research on Cancer says that particulates harbour a known carcinogen, benzoypyrene, which is also present in ordinary soot and even in burnt toast. However, the amounts are so tiny that we need not worry about it—the body's natural defences against cancer-forming agents can easily render it harmless.

The particulate problem is worse in France, where half the new cars run on diesel, and where an estimated 80 000 tons of particulates are emitted each year. By the mid-1990s Peugeot had become the world's biggest diesel-car producer, encouraged by the French government, whose tax policy has diesel costing a third less than conventional petrol. European diesel-car production already exceeds two million annually, and by the year 2000 these cars will be emitting 200 000 tons of particulates a year.

One way to reduce particulate emissions is to trap them in a ceramic filter, and then burn them off. But this wastes fuel. Now the French chemical giant Rhône-Poulenc has come up with another answer: add a little cerium oxide to the fuel. This catalyses the burning of the particulates and eliminates them. Not surprisingly, Rhône-Poulenc are the world's leading suppliers of cerium.

Only a little cerium oxide is needed: 50 g of additive will treat a ton of diesel fuel (about 1400 litres), so that a diesel powered car would require around 1.5 kg of additive during its lifetime. The car could be equipped with a cerium oxide injection cartridge which would last around ten years and cost about £250 ($400).

Cerium is a grey metal which finds little use as such because it tarnishes easily, is attacked by water, and will even ignite if scratched with a knife. It is used in hardening steel, in carbon arcs for studio and flood lighting, and in flints for lighters. Cerium oxide is used for polishing glass surfaces, and is quicker than the traditional iron oxide rouge. The walls of 'self-cleaning' ovens contain cerium oxide, and this too catalyses the oxidation of cooking residues, which are also mainly carbon.

Cerium was first identified by the Swedish chemist Jöns Jacob Berzelius and Wilhelm Hisinger in 1803, who named it after the recently discovered asteroid, Ceres. But it was not until 70 year later that two American chemists, William Hillebrand and Thomas Norton, first obtained a pure specimen of the metal. Cerium is one of the so-called rare-earth groups of elements, which are all so similar that they defied early attempts to separate and purify them.

Cerium is widely dispersed in the environment, and our bodies each contain about 40 mg, but it has no known natural biological role. It is non-toxic, and cerium salts were once prescribed for morning sickness and travel sickness. Because it is so safe, cerium pigments are seen as a potential replacement for toxic metal pigments in paints, inks and plastics. Traffic signs could become much more arresting if they used cerium-based pigments, especially red ones. Red is the colour which has the largest sales, exceeding £500 million ($750 million) per year worldwide.

Cerium sulphide is the non-toxic bright red pigment that is destined to replace those made from the toxic metals, cadmium, mercury and lead. (You will meet two of these metals again in Gallery 8.) Until recently cadmium red was the preferred choice for a red pigment, but stringent regulations are being introduced to restrict cadmium and other heavy metals, such as lead and mercury, which are also toxic. Unfortunately these account for most of the red metal pigments. Cerium sulphide is also a rich red colour and stable up to 350 Celsius. It even creates its own 'colour space' because it produces a different range of reds not obtainable with other pigments. The sulphide is made by heating cerium metal vapour in an atmosphere of sulfur. By adding traces of other rare-earth metals it is possible to produce a range of colours from deep maroons, through brilliant reds, to bright orange.

If these new uses for cerium take off, there could eventually be a demand for this metal of several hundred thousand tons a year. Is there enough cerium ore to supply the world's needs? Rare-earths expert Patrick Maestro, writing in the *Kirk-Othmer Encyclopaedia of Chemical*

Technology, thinks there is, pointing out that cerium is the most abundant rare-earth metal, more common even than copper or lead.

Fuel additives and pigments are not the only reasons why cerium demand will increase. Flat-screen TVs, low-energy light bulbs, and magnetic-optic CDs also use cerium. The country which will benefit most from all this is China, which has the world's largest deposits of cerium-rich ores, with reserves in excess of 36 million tons out of a world total of around 50 million tons. The other deposits are in the USA, India and Australia.

Portrait 9

Spare the trees—calcium magnesium acetate, CMA

It is not only the fuel for our cars and the additives that we put in it which will be a problem for the future. Sometimes we find we cannot use our car because the roads are icy. Keeping city streets clear when winter comes may threaten the plants that border them, and especially trees. Mild winters spare thousands of trees in city streets, not because they are less in danger of being damaged by frost, but because they have not been exposed to the salt which is used to clear snow and ice from roads. A bad winter and heavy salting kills not only hundreds of road-side saplings, but also mature trees.

Planting only salt-resistant trees in city streets and motorway verges is one way of tackling the problem. All trees are affected by salt, but some, like oak and white poplar, are fairly tolerant once they are established. However, this approach is a long-term solution. In future, trees could be saved if city authorities adopted a new environment-friendly de-icer in place of salt. This alternative is calcium magnesium acetate (CMA), developed by chemists in the USA and Europe. It is not only less damaging to plants than salt, but it may actually improve plant growth by making the soil more permeable to air and water. Preliminary results from the USA indicate that CMA causes no measurable damage to trees when it is applied in the concentration at which salt would kill them. In addition, calcium and magnesium are essential nutrients that plants need.

The acetate part of CMA is a bonus as well. This is the negatively charged half of the compound, and unlike chloride, which is the negative half of salt, it does not damage road structures and bridges by attacking the steel reinforcing bars used to strengthen concrete. This advantage of CMA may, in the long run, be the main reason for its introduction, rather than concern over road-side trees and other vegetation. As such it will have to compete with other non-chloride de-icers, such as urea and ethylene glycol. Urea, though cheap, is not very effective, and it can stress the environment in a different way because it acts as a nitrogenous fertilizer. It decomposes to ammonia which is very toxic to fish, and so it can pollute those streams and rivers into which it drains. Ethylene glycol is the anti-freeze used in cars, but unfortunately it is relatively toxic, and so its widespread use would be opposed. It is also slightly slippery, and makes skidding more likely on roads on which it is sprayed. But it is often used to de-ice aircraft wings before take-off. CMA has none of the disadvantages of urea or ethylene glycol, and toxicity tests show that it is less poisonous even than salt.

The chemical principle behind de-icing is that a solution of any chemical always freezes at a lower temperature than water itself. The more soluble a compound, the colder it can get before its solution freezes. A concentrated solution of common salt can remain a liquid down to −21 Celsius. For this reason scattering salt on snow and ice will cause it to melt and drain away.

The great benefit of salt is its cheapness, but repairing the damage caused by it to concrete bridges and elevated motorways can cost a thousand times as much as the salt. When the iron and steel used in road building are exposed to salt solution they rust rapidly. Rusting causes expansion, and cracks the concrete that the steel is meant to be reinforcing. The Thelwall viaduct of the M6 motorway in Cheshire, in north-west England, is an elevated section nearly a mile long. Keeping this free of ice has required about 15 tons of salt in a typical winter at a total cost of about £10,000 ($15,000) during the 25 years of its existence so far. This has so damaged the viaduct that repairs will now cost £10 million ($15 million). More serious was the collapse in 1985 of the Ynysygwas bridge into the river Afan, near Port Talbot, Wales. This was directly attributed to salt attack.

CMA, on the other hand, costs ten or twenty times as much as salt, and even though an application of CMA lasts longer than salt, it is still far from competitive in terms of purchase price. But CMA is an economical proposition for the de-icing of bridges, viaducts and city streets lined with trees because the damage it causes is believed to be negligible. Indeed, tests show that CMA actually protects steel against rusting.

CMA is manufactured from dolomitic limestone, of which large deposits are available. This mineral is a mix of carbonates of calcium and magnesium, which is heated to convert them to their respective oxides, which are then treated with acetic acid to form CMA. So far the cost of CMA has limited its use, and this may well restrain it in future. But perhaps we should be prepared to pay a little extra to protect the trees and plants which share our city environment, and which give us such pleasure in summer.

Portrait 10

Bang! you're not dead!—sodium azide

Of course, if you crash your car into a tree at the side of the road you might take a slightly different view of their aesthetic contribution to the landscape. But as you climb from the wreckage you might appreciate the benefits of the chemical in this final portrait in Gallery 7. Sodium azide appears menacing—after all, it is highly toxic and explosive—but it can spring into action and save your life.

The modern motor car is a triumph of engineering, not least in incorporating safety features designed to protect its occupants in the event of a crash. There are energy-absorbing bumpers (fenders), padded instrument panels, impact beams in the doors, driver and passenger head restraints, seat belts, anti-lock brakes, and strengthened roofs. Despite all this protection people still die in car accidents, and those most at risk are drivers and front-seat passengers. According to the US National Highway Traffic Safety Administration, 20 000 front-seat passenger-car occupants are killed each year, and 300 000 suffer injuries that require hospital treatment.

One way of saving some of these lives is to pack around 250 grams (half a pound) of this rather dangerous chemical, sodium azide, into the car—and to detonate this to inflate airbags if the car crashes. The airbags cushion those in the front seats and prevent them smashing their heads against the steering column, dashboard or windshield. So far airbags have saved more than 1200 lives in the USA alone; meanwhile they have killed around 50 others accidentally by breaking their necks. Most of those people might have died anyway from the other injuries they sustained. Today almost all new cars are fitted with driver and front-passenger airbags. Some vehicles also have side-impact bags to protect against collisions with the sides of the car, which contribute to about a third of passenger fatalities.

Airbags were patented in the 1950s, and were designed to be inflated with gas from a pressurized canister. They failed to catch on because they were unreliable, due to the unpredictable behaviour of the canisters and the variable pressure of the gas in them. The answer was to replace the high-pressure gas canister with a charge of sodium azide, and to generate the gas by 'exploding' this. A measured amount of azide will instantly release exactly the right amount of nitrogen gas.

The sequence of events which triggers an airbag goes as follows. Imagine you are travelling in your car and you collide with another vehicle or solid object. If the impact velocity is above 10 mph, this will register in a sensor in the electronic control unit which decides whether to inflate the airbag. This control unit may be located at the front of the car, or near the foot area, or it may even be packaged with the airbag itself. The sensor analyses the deceleration of the vehicle, and can tell the difference between a life-threatening impact and a bump. If the former, then it activates an initiator (known as a squib) which in turn ignites the sodium azide. This produces a large volume of gas, which is filtered as it inflates the nylon airbag. This cushions the driver and front-seat passenger as they lurch forward, and thereby protects them from fatal injuries. Analysing the impact and inflating the airbag occurs within 25 thousandths of a second, which is five times faster than the blink of an eye. A few thousands of a second later the driver or passenger hits the airbag, which then begins immediately to deflate with the hot nitrogen gas venting sideways in a controlled manner.

In the USA the driver's bag inflates with 70 litres of gas, while the passenger's has around twice this amount since it has to fill a larger volume of space. These airbags are large because they may have to protect people who are not wearing seat-belts. In Europe, where seat-belts are compulsory, the airbags are smaller (about 30 litres for the driver's) and are designed mainly to protect the head and neck against injury.

The mixture of chemicals which works the airbag is called the propellant, and this consists of sodium azide, potassium nitrate and silicon dioxide (silica). The sequence of chemical reactions begins with sodium azide (chemical formula NaN_3) in the squib being ignited by an electrical impulse. This generates a local temperature of 300 Celsius, enough to initiate the rapid decomposition of the bulk of the propellant. First the azide ignites to produce a mixture of molten sodium metal and nitrogen gas. The sodium then reacts with the potassium nitrate to release more nitrogen and form potassium and sodium oxides. These oxides instantly combine with the silica to form harmless sodium silicate glass. Only nitrogen escapes into the air bags.

Sodium azide (NaN_3) is a white crystalline powder consisting of positive sodium ions and negative azide ions. Azide is where the action lies. This strange chemical entity is made up of three nitrogen atoms joined together. Perhaps it is not so surprising that it wishes to revert to the more stable nitrogen gas—nitrogen atoms joined in pairs—and will do so given half a chance. When this happens in a lab the sodium is left behind as sodium metal, and this is one way of producing ultra-pure samples of sodium for research purposes.

Sodium azide is made industrially from sodium amide and nitrous oxide gas, and has a variety of uses apart from airbag propellant. It can be converted to hydrazoic acid and then to other salts such as lead azide, which is used as a detonator. Sodium azide is very toxic, and has been used in agriculture to kill nematodes and weeds, and to control fruit rot. It is also dangerous to humans, and is even more poisonous than cyanide. If its dust is inhaled then it produces severe irritation of the nose, throat and lungs. Azide is a metabolic poison which inhibits enzymes such as cytochrome oxidase and catalase, and has a serious effect on the cardiovascular system leading to high blood pressure, abnormal breathing, irregular heart beats, hypothermia, convulsions, and possibly death. However, it is not carcinogenic.

Clearly it would be better to avoid sodium azide if possible, and in Europe and Asia other explosives, such as aminotetrazoles and aminoguanidine nitrate, are being used to generate gas for airbags. While these produce small amounts of carbon monoxide when they go off, this is not a serious threat to the occupants of the car into which the bag vents its gases.

Airbags solve one problem but they create a problem of their own: their disposal at the end of a car's life. What is to be done with the 100–250 g of sodium azide or other propellant? Controlled incineration is one way to render it harmless, but a better way is proposed: to dispose of them with super-critical water oxidation (see portrait 8 in Gallery 6). which will reduce the propellants to harmless gases such as nitrogen and carbon dioxide.

Another way of getting round the problem of disposal is to replace the propellant with some other source of gas, and there is now a return to the concept which was suggested for the earliest airbags, which is to have a canister of compressed gas such as argon. In so-called 'hybrid' devices this is kicked into action by rapid heating of the argon by a small amount of propellant, causing the canister to burst open and propel its contents into the bag. In many cars the more compact azide-driven airbag is used in the steering wheel, while the passenger airbag is the larger hybrid bag because there is more room on the dashboard to mount it.

The propellant can be done away with altogether if the gas in the canisters contains an explosive mixture, such as hydrogen + air or butane + nitrous oxide. These mixtures can be sparked, and will explode instantly to fill the bag with inert gas. The bags inflate faster than the sodium azide airbags, and this is an advantage for side bags which have less space in which to protect people in a collision.

Popular fears about airbags are that the noise of the explosion will deafen you, and that you may still be knocked unconscious and suffocate with your face buried in the bag. There is no risk of either happening. Most people involved in crashes in which an airbag was deployed did not notice the loud bang, because this was drowned by the noise of the crash itself. Nor are you likely to suffocate, because the airbag is designed to deflate completely within a second of your head striking it.

Almost half of those who find themselves in a collision in which an airbag inflates suffer some form of injury, although generally these injuries are quite minor. A few unfortunate people still suffer concussions and fractures because of the force of the impact. Susan Ferguson of the Insurance Institute of Highway Safety, Arlington, Virginia delivered a careful analysis of fatalities due to airbags at the Airbag 2000 Conference at Karlsruhe, Germany, in November 1996. Those who have died as a result of airbags are mostly either babies or young children and older people who sit too close to the steering wheel and whose heads are violently thrown backwards as the airbag explodes. Babies in rear-facing safety seats in the front passenger seat are most at risk. The energy of a fast-filling airbag can deliver a fatal punch if it strikes at the wrong angle. In one accident, in a shopping centre in Idaho in 1996, a baby girl strapped in a front-facing safety seat was decapitated when her mother's car hit a parked vehicle and the airbag exploded. The horror of such an accident can easily make us question the value of airbags, but it may be the price a few unfortunates have to pay if the rest of us are to be better protected and many lives saved.

GALLERY 8

ELEMENTS FROM HELL

An exhibition of molecules that are mainly malevolent

THE ROAD to hell is paved with good intentions . . . so the old saying goes. In this Gallery I want to show you that this can be indeed true, but it is also true that the road to hell can be paved with evil intentions—sometimes all the way down to the pit of fire. Elements cannot really be described as coming from hell, nor can molecules, but they can produce effects that can only be described as satanic.

Some elements that exist naturally can be very toxic, such as beryllium and lead, and the same is true of some natural molecules, such as atropine. We have seen in other Galleries that when chemists discover a natural molecule which has desirable properties, it is often possible to make a safer version that retains these properties, or even enhances them, while unwanted side-effects can be eliminated or at least toned down. The opposite is also possible. If the desired property of a molecule is its ability to kill, then it is possible to refine that aspect. What was merely dangerous can be made maliciously deadly. We begin our tour of the portraits of Gallery 8 with an inspection of one of these terrible molecules.

■ Portrait 1

The quick and the dead—sarin

Could Adolf Hitler have saved his Third Reich from defeat? Quite possibly. What he needed was a secret weapon to wipe out the Allied troops when they invaded the Normandy beaches of northern France on D-day, 6 June 1944. Then with a quick victory in the west he could have rushed his troops to meet the oncoming onslaught of Russian armies from the east, and maybe even have wiped out those invaders as well.

Hitler was fond of secret weapons. Some, like the jet fighter, the V1 flying bomb and the V2 rocket bomb, were triumphs of engineering and did a lot of damage, but they were generally developed too late to save his empire. In fact Hitler had one secret weapon that was very cheap and easy to make, and that would have stopped advancing armies dead, but he never used it. The weapon was the nerve gas sarin, against which the Allies had no defence because they did not even know of its existence. By 1944 sarin was in production, but the Nazis withheld it in the mistaken belief that the Allies would retaliate with the same chemical.

But what Hitler held back from using in World War II, others have since had no qualms about using to further their ends. Twelve people died and more than 5000 were injured when members of the Aum Supreme Truth doomsday sect released sarin on the Tokyo underground during the morning rush hour on 19 April 1995. An earlier release of sarin by the same group had killed seven people in Matsumoto, Japan, in the previous year.

The Aum Supreme Truth cult's plan was to throw Tokyo into turmoil by releasing sarin on five trains that would converge on the Kasumigaseki subway station at the height of the rush hour at 8.15 am. They knew that many of the staff of the Tokyo Metropolitan Police Department and the National Police Agency used this station and that there would be many policemen and officers on the trains and in the station at that time. The sarin was held in plastic bags inside rolled-up newspapers which were dropped on the floor of the trains by the cult's terrorists, who then punctured the plastic bags with pointed umbrellas as they alighted from their respective trains a few stops before Kasumigaseki.

While pure sarin would have been undetectable by its smell, that used by the Aum cult was only 30% pure and the contaminants were pungent-smelling compounds which alerted passengers to the fact that something was wrong. But as some workers staggered off trains, coughing and collapsing, at stations along the line, other passengers got on. In some coaches passengers panicked as those around them fell to the floor writhing in agony and foaming at the mouth.

Remarkably when the offending package was drawn to the attention of a station master at one of the stations on the Marunouchi Line, he promptly got a broom and dustpan and removed it! The train then continued on its way and for another hour it continued to carry commuters. It not only passed through Kasumigaseki but continued on to the terminus, turned round, and passed through Kasumigaseki again. At the other end of the line, it turned around yet again and eventually passed through Kasumigaseki for a third time, all the while injuring passengers. It was finally halted at 9.27 am.

While the total number of commuters affected reached a final total of over 5500, only 12 died. The 169 hospitals in Tokyo began to fill up with the injured although it was to be two hours before the first correct diagnosis was made—by a military doctor—that the patients were victims of a nerve gas attack. For the next few days and weeks people continued to be admitted to hospital as delayed symptoms of the nerve gas affected them.

In 1988, Kurdish villagers claimed they were victims of a gas attack by the Iraqi government which left many women and children dead. It had also been suspected that Iraq had used sarin in the war with Iran a few years earlier. This was impossible to prove at the time, but it has since been shown that the Kurds' accusations were true. In 1993, soil and shrapnel was collected from bomb craters in Kurdish villages by James Briscoe, an archaeologist working for the US group Physicians for Human Rights, and taken to the British Chemical and Biological Defence Establishment, at Porton Down, England. Chemists there found traces of substances derived from sarin in the soil. They even detected sarin itself on a bomb fragment where it had been protected by being absorbed into the layer of paint. Analytical tests are now sensitive enough to detect less than a billionth of a gram of nerve gas.

Sarin has been described as the poor man's atomic bomb, because of the devastation that a relatively small amount could cause—at least to human lives. Sarin is not a gas but a colourless liquid which boils at 147 Celsius, yet it is volatile enough for its vapour to contaminate the air to lethal levels. It disrupts the central nervous system of those who breathe it in and, unlike the commuters of Tokyo, they would not realize there was a danger because pure sarin does not reveal its presence by an unusual smell. It kills by paralysing the nerves and muscles of the lungs and heart.

Sarin is made from the commercially available chemicals methylphosphonic dichloride, propan-2-ol and sodium fluoride, but sales of these are carefully monitored around the world for just this reason. The danger to would-be manufacturers comes in handling the finished product, because any direct contact with sarin, even a drop of liquid on the skin, can be lethal. As little as one milligram is enough to kill a human. Indeed, an ounce of sarin would be enough to wipe out a town of 25 000 inhabitants, if it were sprayed from the air as a fine mist.

The German chemist Gerhard Schrader discovered the nerve gases tabun and sarin in 1937, when he was working for the chemical combine IG Farben, testing various phosphorus compounds for insecticidal properties. Schrader was not the first to make these kinds of molecules—they had been reported in technical journals in 1902—but he was the first to realize their lethal nature. He also found a better way of making them, and this was patented by IG Farben in 1938. When the Nazis realized just how toxic they were the new chemicals became a closely guarded secret, and were given the code-name N-Stoff. They were tested not only on guinea pigs, but also on apes, and on concentration camp inmates.

In the closing stages of World War II, the staff at IG Farben's headquarters in Frankfurt destroyed all their records, but details of the tests came out during the Nuremburg Trials. In any case, at the end of the war the Allies came across the massive stockpiles of nerve gases. Chemical plants had been making them at the rate of hundreds of tons per month, and there was enough to destroy all human life on Earth.

In his memoirs *Inside the Third Reich*, Albert Speer, who was minister for armaments and war production, says he seriously considered killing Hitler early in 1945 by releasing nerve gas into the ventilation shaft of the Führer's underground bunker in Berlin. However, his plan was thwarted when the inlet shafts were suddenly redesigned to guard against possible gas attacks.

British chemists in World War II were also working on similar phosphorus molecules, but they failed to discover either tabun or sarin. Although they made molecules which were similar, these turned out to be no more toxic than the phosgene and mustard gas which had been used in World War I, so the Allies continued to stockpile these earlier, but less effective, war gases.

Sarin is an organophosphorus compound, which is a term that is often confused with organophosphate, the class of chemicals commonly used as pesticides, such as in sheep dips. Both types of compound are insecticides, and sarin was once tested against the plant-louse phylloxera, which attacks grapevines. A 0.1% solution of sarin was spread on the ground near the vine root and completely eliminated the infestation. Effective as it is, sarin is far too dangerous to use as an insecticide.

The sarin molecule consists of a phosphorus atom bonded to four other atoms or groups of atoms, namely oxygen, fluorine, a propoxy group and a methyl group. It is this last group which determines that the molecule is an organophosphorus compound, because it has a direct carbon-to-phosphorus bond. Organophosphate insecticides lack such a bond, which generally renders them much less toxic to mammals, though they are still deadly to insects.

Sarin is called a nerve gas because it acts in the body to paralyse the central nervous system. It does this by attacking the key enzyme cholinesterase, which is needed to cancel the chemical messenger acetylcholine after it has done its job of transmitting a signal across a nerve junction. An electric impulse travels down a nerve fibre and releases acetylcholine, which triggers an impulse at the next nerve ending. Once it has done its job, the acetylcholine must be removed, and this is what the enzyme does. If the messenger molecule is not cancelled it continues to stimulate the nerve ending unchecked, resulting in twitching, convulsions and possibly death if it is the heart or lung that is affected. Nerve gases block the enzyme so effectively that one tiny drop is lethal.

One of the first symptoms of sarin poisoning is semi-blindness, caused by its effect on the nerves and muscles in the eye. But all is not lost at this stage, and there are now antidotes to sarin which are carried by troops in battle and held in stock at hospitals. Sadly, these antidotes are likely to be needed one day, because sarin will continue to appeal as a weapon of mass terror. Thankfully sarin can be neutralized at relatively little cost, either in antidote treatment, or in clean-up operations. The molecule is easily destroyed by alkaline solutions, and a mixture of washing soda and household bleach will erase it by knocking off its fluorine atom, thereby rendering it non-volatile and non-toxic.

The antidote for nerve gas consists of atropine, which counteracts the effects of the unchecked acetylcholines, together with oxime, which re-releases the sarin-blocked cholinesterase enzyme so that it can function again as normal. Atropine is a curious example of one poison being used to counteract another. Cases of deliberate poisoning by atropine are now rare but not unknown, and the next portrait looks at this natural molecule.

Portrait 2

Fancy a gin and tonic, dear?—atropine

It costs less than £5 ($8) for a gram of atropine, which is enough to poison a wife and lay a false trail of spiked tonic waters on the shelves of a local supermarket. So thought biochemistry lecturer Paul Agutter of Edinburgh, Scotland. His plan almost succeeded when he tried to kill his wife Alexandria in August 1994. By a remarkable coincidence his devious plot was foiled and his wife's life was saved. Agutter was arrested and brought to trial, found guilty of attempted murder, and jailed for 12 years.

The amazing coincidence was that one of the first bottles of Agutter's poisoned tonic ended up in the home of Dr Geoffrey Sharwood-Smith, a consultant anaesthetist who was familiar with the symptoms of atropine poisoning. His wife and son became ill after drinking gin and tonic, and he informed the hospital where they had been taken that he thought they had been poisoned by atropine.

In the next few days five other local people were admitted suffering from atropine symptoms, including Mrs Agutter, the intended victim. Analysis of her gin and tonic proved that it had more atropine in it than the tonic in the bottles in the supermarket, thereby revealing her husband's murderous intentions. His plan was to murder her, inherit her share of the family fortune, and marry his mistress. Agutter had schemed well—his choice of atropine was clever. This poison is metabolized by the body, leaving only tiny traces by the time death occurs. Nor does it act as an irritant, so there are no inflamed internal organs for a pathologist to find when conducting a post-mortem.

In the USA, atropine has caused deaths among teenagers who have tried to get a high by drinking tea made from the leaves of the ornamental bush called angel's trumpet. This plant produces a lot of atropine, which can induce hallucinations in small doses, but too much can cause paralysis and memory loss, and sometimes it kills. The US Center for Disease Control issued a nationwide warning about the epidemic of atropine poisoning from angel's trumpets, and in Maitland, Florida, the city council went even further and made planting it illegal.

A better known natural source of atropine is the deadly nightshade, whose botanical name is *Atropa belladonna* from which the name atropine is derived. One berry of this plant is enough to kill a child—although it rarely does so because its bitter taste immediately acts as a warning and a repellent. The taste of atropine can be detected at concentrations as low as one part in ten thousand, which explained why Agutter disguised it by adding it to tonic water. This essential mixer for gin is already bitter because it contains quinine, and could mask the taste of the atropine.

John Mann, in his fascinating book *Murder, Magic and Medicine*, reports that belladonna was investigated by Cleopatra in her search for the best poison for committing suicide after the defeat of her, and her lover Anthony's, fleet at the battle of Actium in 31BC. She ordered a slave to be given it, and while he died a quick death it was clearly also a painful one. (Further research revealed that the venom of the asp was equally rapid, and led to a relatively tranquil end, so Cleopatra chose that.) Mann speculates that another murderer of antiquity chose atropine: the aristocratic serial-killer Livia, wife of the Roman emperor Augustus. She probably used belladonna to eliminate those who were likely to succeed her husband as emperor, thereby ensuring that her son, Tiberius, would inherit the imperial purple—and he did. We saw a rather unfortunate consequence of this is Gallery 4.

In Renaissance times belladonna became fashionable as an eye cosmetic; indeed the name comes from the Italian for 'beautiful lady.' Women would squeeze the juice of a berry into their eyes and the atropine would cause the pupil to dilate, giving a fashionable doe-eyed look. One treatment would work for several days. Actresses continued to use atropine for this reason even in this century, and so did ophthalmic surgeons when they wished to examine inside a patient's eye.

Atropine is a white, odourless crystalline powder which melts at 114 Celsius, and it was first isolated in 1833 by two German chemists, Geiger and Hess, from the black, shiny, cherry-sized berries of the deadly nightshade. It is still extracted from this tall bush which is native to woodland around the Mediterranean, and which is cultivated in France. As a pure chemical, atropine is not very soluble in water, and doctors who administer it medically choose a derivative such as atropine sulphate, which is very soluble. The amounts given for therapeutic purposes are tiny, and typical doses are less than a milligram. Larger amounts lead to blurred vision, excitement and delirium, but if you intend to kill someone with atropine you will need to give them at least a gram.

Atropine may be a deadly poison, but it is also an antidote for other poisons, such as the carbamate and organophosphate insecticides used in agriculture. As we noted in the previous portrait in this Gallery, it is also an antidote for the deadliest of chemicals, the nerve gases—soldiers in the Gulf War of 1991/2 carried supplies of atropine and pralidoxime to inject themselves in the event of a nerve gas attack. This curious duality of toxin and treatment stems from the effect that atropine has on its target organs, the nerve endings. Atropine immediately calms the nerve endings and the pralidoxime re-releases the enzyme blocked by the nerve gas so that it can begin to do its job again, and normal nerve function is restored.

In the body atropine blocks the production of the messenger molecule, acetylcholine. The first effect is that it dries up bodily fluids like saliva, tears, mucous, phlegm, sweat and urine, and this is why it is given before operations. At various times atropine has been prescribed as a treatment for conditions when there is an excess production of bodily fluids, as in hay-fever, colds and diarrhoea. At one time it was even prescribed as a cure for bed-wetting.

Not every living thing is adversely affected by atropine. The soil bacterium *Pseudomonas putida* actually feeds off the poison, breaking the molecule down to extract its carbon and nitrogen. Because atropine occurs in small amounts in several plants there would be a steady build-up of this natural toxin in the environment were it not for bacteria that digest it.

Portrait 3

The people are revolting—CS gas

Not all chemicals that irritate the body and produce unpleasant symptoms are dangerous toxins. A few, such as CS gas, have been put to use to fight crime, and even used on a wider scale to control angry crowds. While we might almost certainly approve of the one use, we might very much disapprove of the other, especially if our sympathies lie with the protesters. We cannot trace the lineage of CS gas back to an evil dictatorship, as we did with the nerve gases: CS was discovered and developed in a Western democracy.

Police around the world often carry spray cans of CS gas. In fact CS is not a gas but a white solid which melts at 96 Celsius, and the cans contain a solution of CS dissolved in a solvent. CS is not soluble in water, and police sprays contain a 5% solution in methyl isobutyl ketone, which is regarded as a safe solvent. When a jet of CS spray is fired into an attacker's eyes, he or she will immediately be disabled by uncontrollable weeping. CS is regarded as one of the safest ways of incapacitating an aggressive assailant, but it can cause harm.

Alastair Hay of Leeds University, in northern England, specializes in toxicology and is chairman of the UK's Working Party on Chemical and Biological Warfare, which has been monitoring agents like CS for many years. His judgment on CS gas is that in theory it is safe, although those with asthma could react badly to it; and he points out that in South Korea, where it is widely used, people there have been affected for several weeks after exposure to CS.

CS and other eye irritants have been used by riot police for more than 50 years, and are dispersed through a crowd in the form of smoke from canisters, hence the name 'tear gas.' Most were discovered earlier this century as part of military research into chemical warfare agents. The German army was the first to use a tear gas in World War I when they fired shells filled with benzyl bromide at Russian positions near Bolimow on the eastern front, and at French troops near Neuve Chapelle on the western front. Neither the Russians nor the French were aware that they had been attacked in this way, and it is believed that local weather conditions prevented the vapours from dispersing effectively.

During that war more than 20 eye irritants were discovered, and interest in tear gas continued in the post-war years. In 1928 two American chemists, Ben Corson and Roger Stoughton of Middlebury College, Vermont, made a series of new compounds, each with two cyanide groups. While most of these were innocuous materials, they recorded that one had 'disastrous' effects when handled. This was a relatively simple molecule which today we know as CS. It consists of a benzene ring to which is connected a chlorine atom and a carbon-carbon double bond, and it is to one end of this that the two cyanides are attached. Its chemical name is 2-chlorobenzylidene malonitrile. The military gave it the code-name CS as one of a series of what they classed as 'C' agents. Another eye irritant was CN, ω-acetophenone, and this was used as a tear gas until it was discovered to have carcinogenic properties. The worst of the eye irritants is CR, or dibenz-1.4-oxazepine, but this is considered too severe for general use.

All eye irritants act on the sensitive nerve endings of the mucous membrane of the eye by triggering certain key enzymes, which unleashes a flood of tears down the lachrimal duct to wash away the offending molecules. Eye irritants work by attaching themselves to sulphur sites within the enzymes, and it is molecules that can react with these sites which cause the protective response. Only a few molecules of such a material are needed to trigger off the tears.

The enzymes are there to monitor and protect the eyes, and we experience their action when we encounter other so-called lachrymators, some of which are perfectly natural, such as formaldehyde which is present in smoke, and the chemical thiopropanal S-oxide, which chopped onions give off. A growing problem is the lachrymator peroxy-acetyl nitrate, which is what makes summer smog in cities so irritating. All produce the symptoms of enzyme over-activity: a stinging sensation, an immediate closing of the eyelids, a flow of tears and inflammation. Move away from the source of the chemical and within a few minutes these symptoms will disappear, and this is also true for CS, whose effects wear off within about a quarter of an hour. As little as one milligram of CS in a cubic metre of air will incapacitate most people, which is why a tear gas grenade is highly effective at dispersing a crowd. For use against individuals it has to be dissolved and squirted directly at them.

The health and safety aspects of CS were debated for many years, and the British government issued its two-part *Report of the Enquiry into the Medical and Toxicological Aspects of CS* in 1969 and 1971. These confirmed that CS was a suitable agent for riot control because it met the criteria of being effective but harmless. People affected by it quickly recovered, and without the need for medical attention. CS can pose a threat to health, but only at levels several thousand times stronger than that needed for crowd control or in police sprays. Then it may cause serious conditions such as oedema (flooding of the lungs), and this is why a few people have died because of it.

Chemical toxins like sarin, atropine and CS can usually be destroyed fairly easily because they are organic molecules whose toxicity is very much linked to their structure. Change that slightly, and they can be rendered harmless. Metals are different. You cannot destroy a metal atom, and once it has gained access to your body the best you can do is excrete it rapidly, as happens with arsenic or antimony. If that is not possible, then you need to move it to a location where its toxic action is minimal, such as locking it away in your skeleton or your liver. Beryllium, lead and cadmium are three toxic metals that the body finds hard to remove, and yet which threaten it with slow accumulation. The next three portraits are of these metals.

Portrait 4

A novel way to die—beryllium

In 1990 there was an explosion at a Russian military plant in Utika, near the border with China. The blast threw a cloud of dust over the nearby town of Ust-Kamenogorsk, and 120 000 people were exposed to a chemical which causes a lung condition. The chemical was beryllium oxide, and the illness is known as berylliosis.

Victims who are contaminated with excess beryllium suffer inflammation of the lungs, which leaves them breathless. It has long been recognized as an industrial disease among certain metal workers. Happily, few cases are now reported, which is just as well since the condition cannot be cured, although its worst symptoms can be alleviated with steroids. Brief exposure to a lot of beryllium at one go, or exposure to a little over a long period of time, will bring on berylliosis.

The accident in Russia revealed that nuclear weapons were probably being made at Utika. The makers of both nuclear energy and nuclear bombs need a metal which will absorb neutrons, those sub-atomic particles that can split atoms and release the pent-up energy. For the inside of reactors or for the components of nuclear warheads the choice boils down to either beryllium or zirconium. Countries in the West plumped for zirconium, which is not toxic. The Soviets, on the other hand, opted for beryllium—and some have paid a heavy price for choosing this lighter metal.

Forty years ago Isaac Asimov wrote a prophetic short story called 'Sucker Bait', part of a collection entitled *The Martian Way*. A space expedition is sent to investigate a fertile planet where the original colony of settlers all died of a mysterious disease which made breathing progressively difficult, and which killed them all within a few years. The planet had abundant plant life and was apparently ideal for human settlement. So what had happened? Their symptoms suggested a slow poison and yet tests revealed nothing, until it was finally discovered that the planet's soil contained high levels of beryllium.

Beryllium is a rare metal on Earth: the rocks and soil here contain only about 2 ppm, and the oceans even less. A million tons of seawater contains less than a gram. However, there are ores of beryllium which are mined, such as beryl, which is beryllium aluminium silicate. This mineral may be coloured green by traces of chromium and is then known as emerald; a pale blue version is called aquamarine.

Beryllium is valued because it is the only light metal with a high melting point (1278 Celsius). It is not corroded by either air or water, even at red heat. Its alloy with copper has some useful properties, such a high electrical conductivity, and it is widely employed in the petroleum industry for sparkproof tools. Aircraft engineers use it because of its high strength and resistance to wear.

Chemically beryllium is related to magnesium, which, as we saw in Gallery 2, is an essential element of human nutrition. Beryllium can mimic magnesium and displace this metal from certain key enzymes, which then malfunction. Our lungs are particularly sensitive. Workers in industries using beryllium alloys have been most at risk, as have been those making fluorescent tubes which were coated inside with beryllium oxide. But none of us can avoid beryllium altogether, and the average person has around 0.03 mg of beryllium in their body, which is about a millionth of an ounce. This is not enough to affect our health since most of it is stored in our bones.

There is a radioactive form of beryllium, designated beryllium-10 because it has an atomic weight of 10 (made up of 4 protons and 6 neutrons in its nucleus), whereas all other beryllium is beryllium-9 (4 protons and 5 neutrons) and is non-radioactive. Beryllium-10 has a half-life of 1 600 000 years, and is formed by cosmic ray collisions with the upper atmosphere. In 1990 beryllium-10 was detected in Greenland ice cores by Juerg Beer of the Institute for Aquatic Science and Water Pollution in Zurich. He discovered that over the past 200 years, the amount of the radioactive isotope was least when the Sun's activity, as shown by the frequency of sun-spots, was the least. Beer believes that it should be possible to chart a record of the Sun's activity back into pre-history using beryllium in ice cores as a guide, and to link it with other climatic changes.

Portrait 5

Poisoned by stealth—lead

Ingemar Renberg of the University of Umeå, Sweden, specializes in analysing the sediments which settle to the bottom of lakes. Year by year these preserve a record of the dust in the atmosphere which has been washed into the lake by rain and snow, and the record can go back several thousand years. In 1994 Renberg was surprised to find evidence of atmospheric pollution by lead at a time long before the industrial revolution which began around 250 years ago.

Claude Boutron of Domaine University, Grenoble, France, confirmed Renberg's finding by analysing the snow which fell on Greenland, which is also preserved in the ice cores as a record of atmospheric pollutants. He found that the level of lead rose from a natural background level of 0.5 parts per trillion (ppt) to 2 ppt in the first century AD. The Roman Empire was to blame. It was then at its height and expanding, and lead was a valued and vital commodity with production reaching 80 000 tons per year. The chemical symbol for lead is Pb, which comes from the Romans' Latin word for lead, *plumbum*—the same word that also gave us 'plumber' and 'plumb-bob'.

When the Romans conquered Britain in 43AD, they discovered rich lead deposits and started an industry there that was to continue on and off for a thousand years, lasting right up to the Middle Ages. Iain Thornton and John Maskall of Imperial College, London, have been researching lead pollution around these old mines and smelters in Derbyshire and North Wales. They discovered that despite heavy surface contamination around the sites, there is little evidence of lead leaching into surrounding soils or contaminating ground water.

Rome flourished from around 350BC to 400AD, and lead was commonly used for roofs, pipes, cisterns and pewter. Lead is easy to smelt from its ores and easy to work with, because it melts at only 328 Celsius. The Romans also made white lead paint, and used sugar-of-lead syrup to sweeten sauces (sugar-of-lead is the chemical lead acetate). These poisoned the populace, and some believe lead caused the Empire to decline and fall. More likely causes of the decline, which started around AD250, were climate, plague and politics. Around this time the Earth's climate became colder and northern peoples began to move south, putting pressure on the Empire. Plague also appeared and epidemics ravaged the Empire several times. Meanwhile the Empire was split by internal military and religious disputes, and on top of all this was a large imperial bureaucracy. Lead was at most a minor factor in Rome's downfall.

After the Dark Ages, which followed the final fall of Rome in 476AD and lasted for 500 years, lead mining began again in earnest, and people found new outlets for it, such as pottery glazes, bullets and printing type. In the time of the Victorians, lead acetate was used as a medicine for diarrhoea and as hair dye. The Victorians also soldered tins of food with lead, a practice that explained the mysterious disappearance of Sir John Franklin's expedition, which set off in 1848 to search for the North West Passage to the Pacific. When the perfectly preserved bodies of crew members were uncovered in permanently frozen graves in the 1980s, analysis showed they had died of lead poisoning. The cause was proved to be the lead solder of their tinned food—the ratio of lead isotopes in their bodies matched the ratio of lead isotopes in the tins nearby. (The ratio of the isotopes lead-206 to lead-204, varies according to where the lead is mined.)

But it was the twentieth century which saw the biggest contribution to atmospheric lead pollution when the metal was added to petrol. In 1921 Thomas Midgley found that adding tetraethyl lead boosted performance, and by the 1960s all cars were running on leaded petrol. While this may have had a deleterious effect on human health, it had a beneficial effect on engines, and even today when most petrol is sold as unleaded, there is still a small amount of lead in it to protect the engine.

As recorded and preserved in lake sediments and Greenland snow, the atmospheric level of lead shot up and reached a maximum 300 ppt by the late 1970s. Levels are now declining thanks to changes in the formulation of petrol. In 1994 another rather surprising archive of lead levels turned up, this time discovered by Richard Lobinski of the University of Antwerp, Belgium, in the cellars of a wine producer. He analysed Châteauneuf du Pape from a vineyard at the junction of the A7 and A9 motorways in the Rhône region of France and found increasing lead levels in the wine as traffic built up over the years.

Lobinski's findings even reflected the changes in the compound used to make leaded petrol. Tetraethyl lead residues in the wine declined from the 1950s onwards, while tetramethyl lead, which replaced it, increased in those years. Together they reached a maximum 0.5 ppb by 1978. The researchers concluded that if the 1978 wine was drunk regularly it could cause mild lead poisoning, but this is most unlikely to worry anyone because 1978 is one of the best vintages, and the wine costs up to £25 ($40) a bottle. Since 1980 the levels of lead in Châteauneuf du Pape have fallen, and by the mid-1990s they were only one tenth of the level of these earlier years. Old wine may also be contaminated by the lead/tin seal around the neck of the bottle—these were phased out only in the 1980s.

This is not the first time in history that wine has been contaminated with lead. The ancient Greeks used lead to 'sweeten' wine, and although their wine was popular it was also reputed to cause miscarriages. In the Middle Ages vintners adulterated cheap wine with lead, sometimes producing mysterious local outbreaks of stomach cramps, constipation, weariness, anaemia, insanity and lingering death. These are the symptoms of severe lead poisoning. In the eighteenth century in Britain, Devon colic caused many deaths among cider-drinkers until the Queen's physician, George Baker, traced the cause to lead-lined apple presses.

Lead weights were used by fishermen as sinkers, and have also caused lead pollution, in this case leading to the deaths of the many swans that scooped up lost sinkers as they fed in the mud along the bottom of rivers. Lead sinkers have now been supplanted by non-toxic ones.

Curiously, as one form of lead pollution is ended, another appears. In 1994 there was an outbreak of lead poisoning in Hungary, due to red lead being used to colour paprika, the spicy flavouring made from dried red peppers. Red lead is a lead oxide which was used for hundreds of years as a pigment. Eighteen people were arrested, but the number of people they have poisoned will perhaps never be known because Hungarians used the paprika to colour many foods, such as goulash, sausages and salami.

Why is lead so dangerous to health? Most of the lead in our diet passes straight through us, but a little is absorbed into the bloodstream. There it is taken up by the enzymes that make haemoglobin, and it disables them. The result is that a precursor to haemoglobin, called aminolevulinic acid, builds up in the body, and this is what poisons us and causes the toxic symptoms we experience. The gut is paralysed, hence the stomach cramps and constipation and excess fluid in the brain causes headaches and loss of sleep. The reproductive system is affected, which may lead to a miscarriage or even a deformed baby. Anaemia is also a long-term effect.

Inner-city children are thought to be most at risk from lead poisoning, and some environmentalists have blamed lead from traffic fumes for causing the learning difficulties and criminal behaviour that these children exhibit. In the USA lead paint from old housing is also blamed, and there inner-city children are automatically given a blood test for lead before starting school. Both these sources of pollution are now declining markedly, but it remains to be seen if this will lead to better behaved children.

Portrait 6

Running down your batteries—cadmium

We can't easily get rid of the lead we don't want in our body, so we store it in our bones. Cadmium, on the other hand, we store in our liver, and cadmium is even more worrying. Cadmium causes cancer in rats, but not in mice and hamsters. Can it trigger cancer in humans? First indications seemed to say yes, and these were based on epidemiological studies which showed more cancers than expected among those who work with cadmium. But as so often happens, later epidemiological studies by other investigators did not bear out the original alarming findings. George Kazantzis, a professor of occupational medicine in the University of London, carried out a study of 7000 people whose work involves cadmium, but found no link with cancer.

Nevertheless, it is a toxic metal to avoid. In the USA there has been a steady reduction in the permitted levels of cadmium in the air of workplaces, and a European Directive has almost eliminated the cadmium used in plastics.

Yet no one dies of cadmium poisoning, and the reputed 40 mg we each carry in our bodies seems not to affect our health or average life-span. Environmentalists speak of cadmium as the *bête rouge* rather than the *bête noire* of modern living, a reference to the cadmium pigment used to colour plastics bright red. In fact, cadmium pigments can be anything from yellow to brown, depending on the proportions of sulfur and selenium in them. Cadmium used in plastics ends up contaminating the environment as the plastic degrades or is burnt. As we saw in Gallery 7, red pigments in future are likely to be made of cerium compounds.

Some uses of cadmium are going to be hard to replace. For example, in the versatile and rechargeable nickel-cadmium battery the cadmium is used very much to the benefit of the environment by saving natural resources. Such batteries should always be collected for recycling. They are likely to be used more widely in future, and may one day even power city cars.

The Nissan car company's all-electric car, which has a range of 250 kilometres (150 miles), is fitted with a new style nickel-cadmium battery that can be recharged in only 15 minutes. This breakthrough in technology could stimulate demand for the metal from its current world production level of 19 000 tons, of which over half goes into nickel-cadmium batteries, which are 25% cadmium. They are much more efficient for electric vehicles because they are only a third the weight of conventional lead-acid batteries. But it is not weight alone which has brought the breakthrough. Nissan's batteries have very thin plates which allow the heat generated on recharging to dissipate quickly.

Since the 1960s, when alarms were first sounded about cadmium, industry has found alternatives for most products that reach the consumer. According to the industry-funded Cadmium Association, the losses of cadmium to the environment are now negligible. The Association defends the use of cadmium on the grounds that it makes better batteries and more stable plastics. In the aerospace, mining and offshore oil industries cadmium is needed to protect steel, and is better even than zinc for galvanizing this metal.

In motor vehicles cadmium has been phased out, even though it was only used to coat a few of the nuts and bolts that are exposed to the corrosive spray that comes from salted roads in winter. Cadmium is particularly good at protecting steel against attack by the chloride from salt and seawater because the cadmium chloride that forms on its surface is insoluble; zinc coatings, however, form a layer of zinc chloride which is soluble and washes away.

Useful as it may be industrially, cadmium is still regarded by our bodies as a dangerous metal, but because it is a natural part of the environment we have developed ways of coping with it. We could never exclude it completely from our diet even if we wanted to, and we have a weekly intake of about a tenth of milligram, much of this coming from foods such as kidney, shell-fish and rice. Smoking adds greatly to this burden. We absorb some cadmium because it mimics zinc, a metal that is essential to life, as we saw in Gallery 2.

A typical half-pounder hamburger provides 0.03 mg of cadmium, which we can trace back down the food chain to the grass the cattle ate, and ultimately to the soil on which it grew. We may even trace some of the cadmium back to the phosphate fertilizers farmers added to the soil. Phosphate rock from Morocco has over 50 g per ton of cadmium, which is why this fertilizer is no longer permitted in Europe, once its major market. Sewage sludge used as a fertilizer can raise cadmium levels in soils, especially if it comes from industrial areas. The amount of cadmium in most soils rarely exceeds 1 ppm, but there are local 'hot spots' which exceed the 3 ppm limit recommended by the EC to protect soils that are fertilized with sewage sludge. However, some hot spots have levels exceeding 40 ppm. Cadmium contamination comes from three sources: old lead and zinc mines; zinc smelters which often produce cadmium as a by-product; and natural outcrops of certain cadmium-rich minerals, such as the Carboniferous marine black shales. Britain's worst polluted area is around Shipham, in the English county of Somerset, where the spoil from zinc mines that operated up until about 1850 has led to soil levels of 500 ppm, an unenviable world record.

So why don't those living in heavily contaminated areas suffer from cadmium poisoning? The answer lies within the human body. Most cadmium we eat is not absorbed through the gut, and what does get through ends up in the kidneys where it is safely locked away by a protein called metallothionein. This has lots of sulphur atoms that bind to cadmium and immobilize it. Having done this, however, it can't let the cadmium go, with the result that the half-life of this metal in our bodies is about 30 years, meaning that much of what we absorb through our diet is with us for the rest of our life. This is what makes cadmium so insidious, and places it on the UN Environmental Programme's top ten list of hazardous pollutants.

The average adult has around 50 mg of cadmium in their body. Our bodies can go on storing cadmium, but eventually there comes a point when our kidneys can cope no longer. If the level exceeds 200 ppm it prevents reabsorption of proteins, glucose and amino acids, and damages the filtering system which has occasionally led to cases of kidney failure. The price we must pay if we want to enjoy the benefits of cadmium are rigorous controls of its use, and a legal requirement to recycle all spent nickel-cadmium batteries. This way the world would recycle 10 000 tons of cadmium annually, whereas we recycle only a small fraction of this at present.

Portrait 7

A novel way of falling out—thallium

Like selenium, whose portrait is displayed in the exhibition in Gallery 1, the discovery of thallium was also linked to sulfuric acid. Thallium too is deadly if we take in too much, but, unlike selenium, it has no metabolic role. However, it does have an interesting history as a murder weapon.

Thallium was discovered in 1862, and caused an international incident at the London International Exhibition that year. William Crookes, a chemist of the Royal College of Science, discovered the metal when he observed a green flame while testing some impure sulfuric acid. This colour led to the name thallium, from the Greek word *thallos* meaning a green bud. Crookes made several salts of thallium which he exhibited, but he did not manage to isolate the pure metal itself. Claude-Auguste Lamy, a physicist from Lille, France, had achieved this and he put on show a sample of the lead-like metal, and was awarded an Exhibition medal as the discoverer of the element. Crookes was furious, claiming he was the discoverer, which he was. Accusations and counter-accusations flew between London and Paris until eventually the judges awarded Crookes a consolation medal.

Thallium sulphate is a colourless, tasteless salt that can be dissolved in water and given to someone you wish to harm. It isn't a perfect poison, but it comes close. It takes about a week to start working, and when it does it produces symptoms that can easily be confused with diseases such as encephalitis, epilepsy and neuritis. But you are unlikely ever to use or be abused by thallium sulphate, because it is prohibited in most Western countries. It is a different story, however, in Iraq. The Iraqi security forces have used thallium sulphate to dispose of those who were actively opposed to the regime. In the 1980s dissident scientists were poisoned with it, and in 1988 an opponent of the Iraqi regime living in Britain, Abdullah Ali, was killed this way. In 1992 two high-ranking army officers, Abdallah Abdelatif and Abdel al-Masdiwi, fell from favour and quickly fell ill. They escaped to Damascus, were granted emergency visas by the British Foreign Office, and were flown to London where thallium poisoning was diagnosed and treated successfully.

The detective story writer Agatha Christie is often blamed for bringing thallium sulphate to the attention of would-be-poisoners. In 1961 she wrote *The Pale Horse*, in which the effects of thallium poisoning were attributed to black magic curses. She described the symptoms perfectly—lethargy, numbness, blackouts, slurred speech, general debility—but she wasn't the first mystery writer to employ this poison. In

Final Curtain, written in 1947, the novelist Ngaio Marsh had her villain using it even though she had no clear idea of how it worked—she described how those poisoned with thallium dropped dead in minutes. Would-be murderers seeking to emulate her villains would have been very puzzled when their intended victims appear to suffer no ill-effects, although this disappointment might only have lasted a few days.

The most notorious real-life thallium poisoner was the serial killer, Graham Young, who in 1971 put thallium sulphate into his workmates' coffee at a photographic equipment factory at Bovingdon in Hertfordshire, England. He posed as a research chemist and bought the thallium from a chemical suppliers in London. Several workers were taken ill, and two died of the mysterious 'bug'. It was only when Young himself suggested to a visiting health expert that the cause might be thallium that the strange illness was correctly diagnosed. Immediately suspicion fell on Young, and he was arrested when it was discovered that he had a previous conviction for poisoning his relatives. He was found guilty of murder in 1972, and sentenced to life imprisonment. He committed suicide in jail in 1990.

The case was a milestone in forensic detection. At the Metropolitan Police Forensic Laboratories, London, the ashes of Bob Egle, one of Young's cremated victims, were analysed by a technique known as atomic absorption spectrometry. This revealed a level of 5 ppm of thallium, proof that Egle had been poisoned with it.

Thallium was once readily available, and was even part of the medical pharmacopoeia as a pretreatment for ringworm of the scalp. Thallium did not kill the ringworm, but it caused a patient's hair to drop out so that the condition could be more easily treated. This strange effect was discovered by accident about a hundred years ago when thallium was tested on tuberculosis patients as a cure for night sweats. It didn't work, but their hair fell out. A Dr Sabourand, the chief dermatologist at the St Louis Hospital in Paris, reported its depilatory action in 1898, and it became the standard treatment for hair removal for 50 years. Women could even buy thallium compounds over-the-counter in the 1930s as Koremlou cream for removing unwanted hair.

Thallium is not a rare element; it is ten times as abundant as silver. It is spread widely in the environment and detectable amounts find their way into crops such as grapes, sugarbeet and tobacco. About 30 tons of thallium a year are produced as a by-product of lead and zinc smelting. Some of this is used for making special glass for highly refractive lenses, some is used for chemical research, and some ends up as thallium sulfate destined for the Middle East and Third World countries, where it is still permitted and used to kill vermin and other unwanted creatures. In 1976, the year Agatha Christie died, a 19-month-old girl from Qatar was brought to London for hospital treatment for a mysterious disease. The case became famous because a nurse, Marsha Maitland, noticed that the child's symptoms resembled those of victims in *The Pale Horse*, which she was reading. When she reported this to the doctors who were treating the child they immediately tested for thallium, found it, changed the course of treatment and saved the girl's life. Enquiries revealed that her parents had been using thallium sulphate to get rid of cockroaches in their home.

Thallium is an insidious poison because it mimics the essential element potassium. A fatal dose for an adult is around 800 mg (less than a quarter of a teaspoon) and yet doses of 500 mg of thallium salt were prescribed in cases of ringworm. Investigations with radioactive thallium show that it is readily assimilated into the body where it particularly affects potassium-activated enzymes in the brain, muscles and skin, producing the characteristic symptoms. The body is not fooled for long by the thallium it absorbs, and excretes it into the intestines. But this is not particularly effective, since a little further along the gut the thallium is once again mistaken for potassium and reabsorbed. The cure for thallium poisoning needs to break this cycle of excretion and re-absorption, and the best antidote is Prussian blue, the dye of blue ink, which is a complex salt of potassium, iron and cyanide. It was suggested 20 years ago by a German pharmacologist, Horst Heydlauf of Karlsruhe, at a time when thallium poisoning was believed to be incurable. The potassium in the salt exchanges with the thallium, which becomes strongly bound to the dye and is not reabsorbed by the gut.

Thallium was born in controversy and it remains controversial to this day. This metal and its salts are always open to accidental or deliberate abuse—hundreds were affected in Guyana in 1987 and 44 died. They had drunk milk from cows that had eaten molasses poisoned with thallium sulphate that had been intended to kill sugarcane rats.

Portrait 8

Requiem for Mozart—antimony

As for thallium, there is no known biological role for antimony, but we cannot avoid taking some of this element in with our diet. In earlier ages antimony was part of the medicine chest of every doctor as the medicament known as tartar emetic, which was the old name for antimony potassium tartrate. As this name implies, it was used to induce vomiting; indeed, tartar emetic was specifically prescribed for this purpose by doctors. Many heavy drinkers would leave a draught of wine overnight in a goblet made of antimony, to be drunk in the morning. The object was to induce vomiting and cure a hangover. The tartaric and other acids in the wine would have dissolved some of the antimony and produced the emetic. This curious feature of antimony prevented its misuse as a domestic poison, but occasionally it did kill because the medical dose is near the fatal dose, which can be as little as 100 mg.

The golden age of antimony treatments was the eighteenth century, when it was regarded almost as a cure-all. One of its most famous victims may well have been Mozart, who fell ill in the autumn of 1791. On 20 October of that year he told his wife Constanza that he felt he was being poisoned, and he probably was. Although Mozart's rival, Antonio Salieri, confessed to this many years later his testimony is unreliable because he was suffering from senile dementia. Those who believe there was a conspiracy against Mozart, and that he was poisoned, favour mercury as the likely agent. A more credible theory was advanced in 1991 by Ian James of London's Royal Free Hospital. He attributed the cause of death to antimony, a poison that Mozart may have been given by his doctor—not to kill him but to cure him.

The facts are briefly these: that autumn Mozart was probably suffering from severe depression exacerbated by debt, over-work and a mauling by the critics, who did not like his new work *La Clemenza di Tito*. He had also received a commission from a mysterious stranger to write a requiem, and as the autumn progressed he became obsessed with the idea that he was writing this for his own funeral. He believed, and perhaps rightly so, that there really was a conspiracy against him. Mozart was a hypochondriac, and regularly dosed himself with a variety of medicines, so much so that he had run up a drugs bill with the apothecaries of Vienna to the tune of £2,000 ($3,000) in today's money. On 20 November Mozart suddenly developed a fever; his hands, feet and stomach became swollen, and he had attacks of vomiting. A second opinion was sought and an illness we no longer recognize, miliary fever, was diagnosed. He died of this curious disease on 5 December, and being virtually penniless he was buried in an unmarked pauper's grave. The conspirators, if they really were a Salieri-led clique of envious court musicians, could have wished for nothing more—except perhaps that his music could have been buried with him.

Ian James believes that Mozart's illness was in fact antimony poisoning since the symptoms are identical. Antimony was prescribed for what in those days was called melancholia. In small doses it leads to headaches, fainting and depression. In large doses it can prove fatal within days, even though there is rapid and continued vomiting. Doctors, and especially vets, prescribed antimony salts well into the following century, and some famous Victorian murderers used these to dispose of unwanted partners—the symptoms were seen as due to normal gastric disorders. Severin Klosowski, alias George Chapman, killed three of his partners this way, and was executed in 1902.

Antimony salts were used until fairly recently against the tropical parasitic infection shismatosis. The antimony atom attaches itself to the sulfur atoms at the active site of certain enzymes, and if these are more important to a parasite than to the human host, then it is possible to kill the one without harming the other too much. Too much antimony, though, and even the human enzyme system can be so disrupted that death eventually ensues, perhaps after several days. Our daily dietary intake is variable depending on what we eat, but is probably 0.5 mg or less, and the total weight of antimony in the average person is around 2 mg. Luckily our body excretes it rapidly and there is no organ which harbours it.

Annual production of antimony is around 50 000 tons per year. It comes mainly from China, Russia, Bolivia and South Africa, where deposits of antimony sulfide ores are to be found, but it is also obtained as a by-product from some copper ores.

Antimony is one of a limited number of metals and alloys which expand on cooling and solidifying. It was known to ancient civilizations 5000 years ago, and craftsmen then used this curious property to make finely cast objects. Antimony was rediscovered by an unknown alchemist in the Middle Ages, and put to use in making alloys for bells, and later for casting lead alloy for type. A little antimony not only hardened the lead, but caused it to expand as it cooled, which gave greater definition to the casting and a clean typeface. Even today its alloy with lead is important, and accounts for most of its use as a hardener for lead plates in car batteries. Research into antimony still continues, with research chemists using it to make new semiconductors such as gallium arsenide antimonide.

The other main use of antimony is as antimony oxide which is added to plastics to make them fire-retardant. When the plastic starts to burn it reacts with the antimony oxide to generate a chemical film that delays burning and may even extinguish the fire. In this way the foam in furniture and mattresses can be made safer, but this had a curious outcome in the UK in 1994. When the bodies of babies who had died of cot death were analysed it was claimed that they contained much higher levels of antimony than normal babies, and that this antimony was coming from their mattresses in the form of volatile gases, which can be formed by bacteria. It was suggested that the babies had died as a result of inhaling these antimony gases, which are indeed deadly at high concentrations. A classic media scare was under way, and tens of thousands of parents threw their cot mattresses away. And all to no effect—because the fault was not with the mattresses, but with the chemical analyses of the babies' organs and the interpretation of what these figures meant. None of the babies who had died of cot death had died as a result of the minute traces of antimony in their bodies.

Portrait 9

Polluting the planet—plutonium

In August 1945, a new artificial element, plutonium, burst on an unsuspecting world when it detonated over Nagasaki, Japan, killing 70 000 people. (The earlier bomb on Hiroshima was a uranium bomb.) Today there exists about 1200 tons of plutonium, of which 200 tons have been made for bombs, and the rest has accumulated as a by-product of the nuclear power industry. It seems that the world is now destined to live with this unwanted metal for hundreds of thousands of years.

Glenn Seaborg, Arthur Wahl and Joseph Kennedy were the first to make atoms of plutonium, in December 1940 at Berkeley, California, by bombarding uranium oxide with deuterium. They christened it plutonium after the outermost planet of the solar system, Pluto. When Seaborg and colleagues investigated their new element they quickly realized that they had stumbled upon a remarkable metal. Its most compelling property was that it was fissile—in other words when an atom of plutonium was hit by a neutron, it split, releasing a lot of energy and expelling more neutrons. These in turn could split more atoms, starting a chain reaction which, given a certain minimum amount of metal, could end in an explosion. This minimum, the so-called critical mass, turned out to be a surprisingly small 4 kg, about the size of an apple.

Within a year Seaborg's group had made enough plutonium for them to be able to see it, and not just detect its presence by its radioactivity. By the end of 1941 they had made sufficient plutonium to weigh a piece, although it weighed a mere three millionths of a gram. If only they had stopped at that point. But by the summer of 1945 enough plutonium had been made for two atomic bombs, the first of which was tested at Alamogordo, New Mexico, in July. So began the contamination of the planet with this least-loved of all the elements.

Only about a quarter of the plutonium in an atomic bomb explodes: the rest vaporizes. The same is true of hydrogen bombs, which have an initiating plutonium bomb at their core. Consequently during the 1950s, when many such bombs were tested above ground, enough plutonium was scattered to the winds to ensure that we each now have a few thousand atoms in our body.

Plutonium is dangerous because it tends to concentrate on the surface of our bones rather than being uniformly distributed throughout the bone mass like other heavy metals. For this reason permissible body levels of plutonium are the lowest for any radioactive element. It decays by emitting α-rays, which are feeble enough to be stopped by a sheet of paper or even by skin, but inside our body these rays can damage DNA and possibly start a cancer such as leukemia. Clearly it is not an element that we want to see loose in the environment.

Encased in a steel can or even a plastic bag, a small piece of plutonium is safe to handle, and it is possible to feel that it is permanently warm due to its radioactivity. This heat can be used to generate electricity, and plutonium units were taken on the Apollo space flights and left on the Moon to power seismic devices. Plutonium has also been used as power sources in deep-sea diving suits and in heart pacemakers. Some plutonium is used to make the heavier neutron-emitting element californium, which is used in cancer therapy, moisture gauges, and on-the-spot analysing equipment for gold prospecting and oil-well drilling.

Plutonium has a density of 20 kg per litre, slightly higher than gold, and it melts at 641 Celsius. The metal is unusual in that it can exist in six different forms, and will change under its own internal heat. As it gets near to its melting point it actually shrinks as it converts from one form to another. Plutonium is unlike other metals in that it is a relatively poor conductor of heat and electricity. The pure metal is as brittle as cast iron, but alloyed with 1% aluminium it becomes as soft as copper.

Plutonium is chemically very reactive. It combines with oxygen to form the oxide, which is potentially very dangerous, as scientists at Los Alamos National Laboratory, New Mexico, discovered in 1993. A canister containing the metal which had not been made air-tight split open under the pressure of the oxide, which is 40% larger in volume than the metal itself.

Accidents like this worry us because unwanted plutonium will have to be stored safely for hundreds of thousands of years—its half-life is

24 100 years. The favoured approach is to bury it in the form of glass logs. The Americans, who are collecting plutonium as they decommission their own and Russia's nuclear weapons, plan to fuse plutonium oxide with the oxides of silicon, boron and gadolinium to turn it into glass. The boron and gadolinium in the glass will ensure that any neutrons are safely absorbed. Nor need we fear that the logs might slowly be attacked by water, which might leach out the plutonium. Plutonium oxide is one of the least soluble of all oxides—it takes a million litres of water to dissolve one *atom*—and the plutonium oxide which is trapped in the glass is even less soluble than this,

Portrait 10

Radioactive lifesaver—americium

In 1945, in addition to plutonium, another new element was announced, and in rather an unusual way. This was americium, and what made it newsworthy at the time was the medium chosen to publish the information: a US children's radio show called Quiz Kids. The guest scientist on the panel that week was Glenn Seaborg, whom we met in the previous portrait. Seaborg had been heading a research programme that had been secretly making new elements at the University of Chicago, Illinois, during World War II. Americium is element number 95, and it had been made by bombarding plutonium with neutrons in a nuclear reactor.

Seaborg announced the formation of his new element, but did not name it. That happened the following year on 10 April at a meeting in Atlantic City, when it was announced that it would be called after America. It came below europium in the periodic table, and this had been named after the continent on which it had been discovered, by E.A. Demarçay in Paris in 1901. So Seaborg and his co-workers decided to call their new element after the continent of its birth, and so we got americium.

By then some of its properties had been measured, and we now know that americium is a silvery, shiny metal which is attacked by air, steam and acids. It is denser than lead and melts at 994 Celsius. Several americium compounds have been made, and some, such as americium chloride, are coloured a pretty pink.

Americium is highly radioactive, emitting α-particles and γ-rays as it transmutes to neptunium. Two isotopes are now produced: americium-243 is made in kilogram quantities by neutron bombardment of plutonium-239, and it has a relatively long half-life of 7370 years; in other words, during this time half of the sample will undergo radioactive decay. The other isotope is americium-241, which is extracted from nuclear reactors, and this has a half-life of 432 years. Americium is potentially dangerous because of its radiation, and if it gets into the body it tends to concentrate in the skeleton. But despite its radioactivity, americium actually saved lives, and several kilograms are made each year.

Today radioactive americium is to be found in most homes, because it is an essential part of smoke detectors. The metal is there to ionize molecules of the air in the detector. Ions are formed by the α-particles ejected by americium as it disintegrates. When an α-particle collides with a molecule of the air, such as oxygen or nitrogen, it knocks off a negative electron and leaves the molecule as a positive ion. This is attracted to the negative electrode in the detector, and as it moves, carrying its positive charge, it creates a small current of electricity. The steady flow of this tiny current is monitored by the electronic circuitry of the detector, and if it is interrupted then an alarm sounds. Smoke disrupts the current because soot particles in the air absorb the ions.

The mount of radioactive americium-241 in a smoke detector is tiny, and most americium sources for smoke detectors are produced in the USA as gold-covered sealed foils, each containing about 150 mcg of americium oxide. They are reliable and cheap, which is why this type of smoke detector is the most popular. The US Atomic Energy Commission first offered americium oxide for sale in March 1962, at a price of $1,500 per gram, and this is still the price today. A gram will supply enough of the oxide for more than six thousand detectors. Smoke detectors in the UK must conform to standards set out by the National Radiological Protection Board, although the Board admits that disposal is not supervised. But it really is not necessary to worry about this, because the radiation hazard which americium poses is insignificant compared to the benefit which it confers by saving lives. Although 33 000 americium atoms per second undergo radioactive decay in a smoke detector, no α-particles escape because these cannot penetrate the wall of the container. Even in air they barely get more than a few inches before they collide with oxygen or nitrogen molecules, and as they do, the α-particlers steal electrons and become atoms of helium gas.

Americium can also be made to produce neutrons, and these can be used in analytical probes. When an α-particle hits a beryllium atom it transmutes this element into carbon, and as it does so it emits a neutron. The stream of neutrons from such americium-beryllium sources can be used to test the metal containers designed to hold radioactive materials, to ensure they are completely radiation-proof.

The γ-rays that americium gives off have shorter wavelengths than X-rays, and so are more penetrating. They were once used in radiography to determine the mineral content of bones and the fat content of soft tissue, but are now only used to determine the thickness of plate glass and metal sheeting. The transmission of the rays tells us how thick the material is.

There are no natural sources of americium, and it is unlikely that any has existed on Earth. Even if there had been americium on the planet when it first formed, it would not have lasted long. The more stable isotope is americium-243, which has a half-life of 7370 years. If there had been a *billion* tons of americium when the Earth came into being this would have been reduced to a single atom in less than a million years.

Americium has no role to play in living things, and it is highly dangerous because of its α-radiation. Just as we saw with plutonium, these rays can be stopped by a sheet of paper, but if they are generated within the body they can wreak havoc on nearby cells. Most americium tends to concentrate in the skeleton, but thankfully we are unlikely ever to ingest any. We might worry about throwing old smoke detectors away, and if they end up being incinerated they add to the atmospheric level of radioactivity, but their release of americium is not going to add significantly to the burden of background radioactivity which already exists on this naturally radioactive planet.

Portrait 11

Land ahoy!—element 114

An unexpected by-product of the development of nuclear weapons during World War II, and of the nuclear power programme that followed it, was a series of new, heavy, artificial elements. Uranium, element 92, was not the ultimate element in the natural order of things, as earlier chemists had believed, although this was the heaviest element found on Earth. The periodic table of the elements could now be extended beyond uranium, but there was no evidence that the new elements had ever existed in nature, and even if they had they would by now have all decayed away.

This decay would have happened if their half-lives were short compared to the age of the planet, which is about 4.5 billion years. We now know that even if there had been a million tons of neptunium, the first element beyond uranium, when the Earth was formed it would still all have vanished, and it has a relatively long half-life of 2 140 000 years. What this means is that in this length of time a million tons would become half a million tons, and in a further 2 140 000 years it would be down to a quarter of a million tons, and so on. In the life-time of the Earth there would have been time for this halving process to happen 2000 times. In fact it would require only 91 half lives, taking 195 000 000 years, to reduce a million tons of neptunium to a single atom. Even if there had been a *million million* tons of neptunium when the Earth was formed, it would still all have decayed away by now, requiring around 114 half-lives, and taking about 250 million years to do so.

Elements beyond uranium have been made in two ways. Some have been made by bombarding an existing element with neutrons, which are absorbed by the nucleus relatively easily because they are not charged and so not repelled. However, absorbing a neutron does not in itself create a new element; but the nucleus then ejects a negative β-particle and a new element is formed with a higher positive atomic number. This is how neptunium (93), americium (95), einsteinium (99) and fermium (100) were made.

Another method of making a new element is to bombard a target of a heavy element with the nucleus of another element, such as hydrogen (atomic number 1), helium (2), carbon (6), nitrogen (7) or oxygen (8), hoping thereby to fuse the two nuclei together and so create a new heavier element. The heaviest elements are synthesized this way. The difficulty here is that both target and missile are positively charged, and so they repel each other. If a new, merged, nucleus is to form then the missile nucleus must have a very high energy. This can be provided by accelerating the nuclei in machines called cyclotrons running at high voltages. In this way other heavy elements were made.

For thirty years the Americans were at the forefront of this research, and the scientists most closely associated with the discovery of many new elements were Albert Ghiorso, who headed the team of physicists and chemists at the Berkeley labs, and his colleague Glenn Seaborg, who was instrumental in making many of them. To begin with the Americans named the new elements after outer planets of the Solar system, just as uranium has been named after Uranus way back in 1789. So we got neptunium (93), and plutonium (94) whose portrait you were viewing earlier in this Gallery.

At element 95 the supply of names derived from the outer planets ran out, so it was not surprising that the scientists turned nearer to home and called the new elements americium (95), berkelium (97) and californium (98)—these last two were named after the laboratories at Berkeley, California, where they were made.

Curium (96) was named after the husband and wife team Pierre and Marie Curie. Einsteinium (99) was a tribute to Albert Einstein, and fermium (100) a tribute to Enrico Fermi, the Italian physicist who won the Nobel prize in 1938 for his pioneering work in this area. Mendelevium (101) was called after the Russian chemist Dimitri Mendeleyev, who produced the first periodic table of the elements in 1869, and nobelium (102) after Alfred Nobel who invented dynamite and used the income it generated to finance the world's most coveted prizes. Lawrencium (103) was in honour of Ernest Lawrence, the man who won the 1939 Nobel prize in physics and who had invented the cyclotron with which many of these new elements had been made.

Elements beyond fermium (100) have very unstable nuclei, and it becomes progressively harder to make and detect them. As a result there were conflicting claims to have discovered elements 104, 105 and 106. In 1964 Russian scientists at Dubna announced element 104, but this was disputed in 1969 by the Berkeley scientists who reported element 104 and deftly named it rutherfordium after Lord Rutherford, the New Zealander who was the first to split the atom. Unlike the Russians, the Americans had managed to make several thousand atoms of element 104.

Element 105 likewise ran into trouble. It was first reported by the scientists at Dubna in 1967, but again the claim was disputed by the Berkeley group, who in 1970 reported it and named it hahnium after Otto Hahn, the German chemist who first observed uranium fission. Several atoms of element 105 have since been made from californium by bombarding it with nitrogen nuclei. In 1997 its name was changed to dubnium in recognition of Russian achievements in this field.

The same clash of claim and counter-claim followed for element 106, but it was almost certainly first produced in 1974 by teams at Berkeley and the Lawrence Livermore National Laboratory, and in 1994 they gave it the name seaborgium in honour of Glenn Seaborg. Several atoms of seaborgium had been made by bombarding californium with oxygen using an 88-inch-diameter cyclotron which produces about a billion atoms per hour of which only one is seaborgium.

Then both the Americans and the Russians were upstaged as element-makers by a new group at the nuclear research facility Gesellschaft für Schwerionenforschung (GSI) in Darmstadt, Germany, led by Peter Armbruster and Gottfried Münzenber. They reported element 107 in 1981. The Darmstadt group used the so-called 'cold fusion' method in which a target of bismuth was bombarded with atoms of chromium. A single atom of the element 107 was detected, and it has been named bohrium in honour of the great Danish physicist, Niels Bohr, who proposed the first successful theory to explain the nature of atoms, and who won the Nobel Prize for Physics in 1922.

Element 109 was discovered in 1982 at the GSI, and in fact was found before element 108, which was reported in 1984. Again the scientists used the cold fusion method, in which a target of lead was bombarded with iron to give element 108. They named this hessium after the German state of Hesse where it was first made, and they called element 109 meitnerium after the Austrian physicist Lise Meitner, who was the first scientist to realize that spontaneous nuclear fission was possible. A single atom of meitnerium was made by the fusion of bismuth and iron. To date fewer than ten atoms of this element have been produced.

Late in 1994 the German scientists made elements 110 and 111, so far unnamed. Again only a single atom of element 110 was made by fusing nickel and lead atoms. Its half-life was 107 microseconds, and it decayed by emitting an α-particle rather than undergoing nuclear fission. A single atom of element 111 was made similarly. Each new element that was produced had a half-life shorter than the element before it in the periodic table.

In February 1996 the German team made the first atoms of element 112, and discovered that the half-life of this element was slightly longer than expected, although it still existed for less than a millionth of a second. The increased nuclear stability for this heavier element is not as strange as it might appear. Several years ago it was suggested that as the atomic number increased we would reach what was termed the 'island of stability' centred on element 114, because this should have a particularly stable nucleus. Elements adjacent to 114 would also have enhanced stability of the nucleus, so that elements 112, 113, 115 and 116 might all be stable enough for several atoms of them to be collected. With their new element 112, the German explorers could be glimpsing this fabled land.

What might they find? It is just conceivable that element 114 will have a half-life long enough for it to exist for a few hours, in which case it may even be possible to study some of its chemical compounds, and thereby show that it is similar to lead. Why lead? This is the element below which element 114 will sit in the periodic table of the elements, and so it will have similar properties. For example, we can predict that it will have two chloride compounds, one with two and the other with four chlorine atoms.

It is most unlikely that these compounds would be stable, because they would suffer from the intense radioactivity associated with these elements, and fly apart as soon as they were made. They might indeed be inhabitants of the island of stability, but this is only a relative term, comparing them to the elements which come before. In reality, they will still be elements from hell.

Happily, we inhabit a planet where most of the radioactive isotopes that were around when it formed have long since decayed away. And yet there is a natural background radiation against which we live our lives. The Earth's crust contains radioactive elements, such as thorium and uranium, and there are a few elements with long-lived radioactive isotopes, such as potassium, but the residue of these is now tiny. Radioactive carbon-14 is formed in the Earth's atmosphere and will always be with us—there is a trace of it in every living thing. There is nothing we can do about any of these. In the past we have increased the number of radioactive isotopes and elements in the environment through nuclear bombs, nuclear testing and nuclear accidents. Compared to these the amount of radioactive elements used for medical and analytical purposes and for smoke detectors is tiny, and their contribution to the background radiation in our lives is very small.

BOOK LIST

If you would like to know more about the chemistry of the molecules and elements in *Molecules at an Exhibition*, then the following reference books are recommended.

H.-D. BELITZ and W. GROSCH, *Food Chemistry*, Springer, Verlag, Berlin, 1987.

S. BUDAVARI (editor), *The Merck Index*, 11th edition, Merck, Rahway, N.J. 1989.

H. G. ELIAS, *Mega Molecules*, Springer Verlag, Berlin, 1985.

J. EMSLEY, *The Elements*, 3rd edition, Oxford University Press, Oxford, 1998.

N. N. GREENWOOD and A. EARNSHAW, *Chemistry of the Elements*, Pergamon Press, Oxford, 1984.

G. M. LOUDON, *Organic Chemistry*, 2nd edition, Benjamin/Cummings, Menlo Park, CA, 1988.

For those readers who would like to pursue some of the topics in a little more depth, the following books are recommended.

A. ALBERT, *Xenobiosis, Food, Drugs and Poisons in the Human Body*, Chapman and Hall, London, 1987.

M. ALLABY, *Facing the Future*, Bloomsbury, London, 1995.

P. W. ATKINS, *Molecules*, Scientific American Library, New York, 1987.

P. W. ATKINS, *The Periodic Kingdom*, Weidenfeld & Nicolson, London, 1995.

A. E. BENDER, *Health or Hoax?* Sphere Books, London, 1986.

S. BINGHAM, *The Everyman Companion to Food and Nutrition*, J.M. Dent, London, 1987.

S. BRAUN, *Buzz: the science and lore of alcohol and caffeine*, Oxford University Press, New York, 1996.

W. H. BROCK, *The Fontana History of Chemistry*, Fontana Press, London, 1992.

A. R. BUTLER, 'Pass the rhubarb', *Chemistry in Britain*, June, **461**, 1995.

C. COADY, *Chocolate, the Food of the Gods*, Pavilion Books, London, 1993.

P. A. COX, *The Elements: Their Origin, Abundance and Distribution*, Oxford University Press, Oxford, 1989.

P. A. COX, *The Elements on Earth*, Oxford University Press, Oxford, 1995.

T. P. COULTATE, *Food, the Chemistry of its Components*, 3rd edition, Royal Society of Chemistry, London, 1995.

H. D. CRONE, *Chemicals & Society*, Cambridge University Press, Cambridge, 1986.

J. DAVIES and J. DICKERSON, *Nutrient Content of Food Portions*, Royal Society of Chemistry, London, 1991.

B. DIXON, *Power Unseen*, W.H. Freeman/Spektrum, Oxford, 1994.

H. G. ELIAS, *Mega Molecules*, Springer Verlag, Berlin, 1985.

H. B. GRAY, J. D. SIMON and W. C. TROGLER, *Braving the Elements*, University Science Books, Sausalito, CA, 1995.

J. EMSLEY, *The Consumer's Good Chemical Guide*, Spektrum, Oxford, 1994.

G. HAISLIP, 'Chemicals in the Drug Traffic', *Chemistry & Industry*, 20 September, **704**, 1993.

C. A. HEATON (editor), *The Chemical Industry*, Blackie, Glasgow, 1986.

M. HENDERSON (editor), *Living with Risk*, British Medical Association/John Wiley, Chichester, 1987.

H. HOBHOUSE, *Seeds of Change: Five Plants that Transformed Mankind*, Sidgwick & Jackson, London, 1985.

J. T. HUGHES, *Aluminium and Your Health*, Rimes House, Cirencester, UK, 1992.

R. J. KUTSKY, *Handbook of Vitamins, Minerals and Hormones*, 2nd edition, Van Nostrand Reinhold, New York, 1981.

J. LENIHAN, *The Crumbs of Creation*, Adam Hilger, Bristol, 1988.

J. MANN, *Murder, Magic and Medicine*, Oxford University Press, Oxford, 1994.

J. L. MEIKLE, *American Plastic: a Cultural History*, Rutgers University Press, New Brunswick, NJ, 1995.

P. J. T. MORRIS, *Polymer Pioneers*, The Center for History of Chemistry, publication no. 5, Philadelphia, 1986.

S. T. I. MOSSMAN and P. J. T. MORRIS (editors), *Development of Plastics*, Royal Society of Chemistry, London, 1994.

M. A. OTTOBONI, *The Dose Makes the Poison*, 2nd edition, Van Nostrand Reinhold, New York, 1991.

J. POSTGATE, *Microbes and Man*, 3rd edition, Penguin Books, London, 1992.

J. V. RODRICKS, *Calculated Risks, Understanding the Toxicity and Health Risks of Chemicals in Our Environment*, Cambridge University Press, 1992.

B. SELINGER, *Chemistry in the Market Place*, 4th edition, Harcourt Brace Jovanovich, Sydney Australia, 1988.

N. M. SENOZAN and J. A. DEVORE, 'Carbon Monoxide Poisoning', *Journal of Chemical Education* 1996, volume 73, number 8 (August), page 767.

C. H. SNYDER, *The Extraordinary Chemistry of Ordinary Things*, John Wiley, New York, 1992.

J. A. TIMBRELL, *Introduction to Toxicology*, Taylor & Francis, London, 1989.

N. J. TRAVIS and E. J. COCKS, *The Tincal Trail: a History of Borax*, Harrap, London, 1984.

D. M. WEATHERALL, *In Search of a Cure, a History of Pharmaceutical Discovery*, Oxford University Press, Oxford, 1990.

E. M. WHELAN, *Toxic Terror*, Prometheus Books, Buffalo, NY, 1993.

G. WINGER, F. G. HOFMANN and J. H. WOODS, *A Handbook on Drug and Alcohol Abuse, Biomedical Aspects*, 3rd edition, Oxford University Press, Oxford, 1992.

APPENDIX

Molecular portraits on display in *Molecules at an Exhibition*

Gallery 1

Phenylethylamine
Oxalic acid
Caffeine
Phosphoric acid
Dipropenyl disulfide
Methyl mercaptan
Selenium
Salicylates
Phthalates

Gallery 2

Calcium phosphate
Sodium chloride
Potassium chloride
Iron
Magnesium
Zinc
Copper
Tin, vanadium, chromium, manganese, molybdenum, cobalt and nickel

Gallery 3

Folic acid
Arachidonic acid
Nitric oxide
Keratin
Mistletoe
Penicillin

Gallery 3A

Ecstasy
Cocaine, heroin and designer drugs
Nicotine
Epibatidine
Melatonin

Gallery 4

Surfactants
Phosphates
Perfluoropolyethers
Sodium hypochlorite
Glass
Ethyl acrylate
Maleic anhydride
Carbon monoxide
Bitrex
Zirconium
Titanium

Gallery 5

Tencel
Celluloid
Ethylene
Polypropylene

Teflon
Poly(ethylene terephthalate)
Polyurethane
Polystyrene
Kevlar

Gallery 6

Oxygen
Nitrogen
Argon
Ozone
Sulfur dioxide
DDT
Dichloromethane
Water
Aluminium sulfate

Gallery 7

Carbon
Ethanol
Methanol
Rape methyl ester, RME
Hydrogen
Methane
Benzene
Cerium
Calcium magnesium acetate, CMA
Sodium azide

Gallery 8

Sarin
Atropine
CS gas
Beryllium
Lead
Cadmium
Thallium
Antimony
Plutonium
Americium
Element 114

INDEX